JUV/
QC
981
.A49
2002

SOCHGO

The Facts on File weather and climate ha

P9-CMO-273

Chicago Public Library

CPL

REFERENCE

Form 178 rev. 11-00

10/04 $ 35.00

CHICAGO PUBLIC LIBRARY
SOUTH CHICAGO BRANCH
9055 S. HOUSTON AVE. 60617

THE FACTS ON FILE
WEATHER AND CLIMATE
HANDBOOK

THE FACTS ON FILE
WEATHER AND CLIMATE HANDBOOK

MICHAEL ALLABY

CHICAGO PUBLIC LIBRARY
SOUTH CHICAGO BRANCH
9055 S. HOUSTON AVE. 60617

Facts On File, Inc.

The Facts On File Weather and Climate Handbook

Copyright © 2002 by Michael Allaby

All rights reserved. No part of this book may be reproduced or utilized in any form or by any means, electronic or mechanical, including photocopying, recording, or by any information storage or retrieval systems, without permission in writing from the publisher. For information contact:

Facts On File, Inc.
132 West 31st Street
New York NY 10001

Library of Congress Cataloging-in-Publication Data

Allaby, Michael.
     The Facts on File weather and climate handbook/Michael Allaby.
          p. cm.
     Includes bibliographical references and index.
     ISBN 0-8160-4517-8
     1. Climatology. 2. Weather. I. Title.
   QC981 .A49 2002
551.6—dc21
2001050114

Facts On File books are available at special discounts when purchased in bulk quantities for businesses, associations, institutions, or sales promotions. Please call our Special Sales Department in New York at 212/967–8800 or 800/322–8755.

You can find Facts On File on the World Wide Web at
http://www.factsonfile.com

Cover design by Cathy Rincon
Illustrations by Richard Garratt

Printed in the United States of America

VB Hermitage 10 9 8 7 6 5 4 3 2 1

This book is printed on acid-free paper.

SOC
$35.00

R0402263056

For Ailsa
— M. A.

To my late wife, Jen, who gave me inspiration and support for almost 30 years
— R. G.

CHICAGO PUBLIC LIBRARY
SOUTH CHICAGO BRANCH
9055 S. HOUSTON AVE. 60617

CONTENTS

ACKNOWLEDGMENTS

I first studied meteorology very many years ago, during the time I spent training as a pilot. I became seriously interested in the subject, and also in climatology, more recently when Jim Lovelock and I collaborated in writing several articles and two books, all of which were directly or indirectly about atmospheric science. My interest continued and I am grateful to Jim for all that I learned about the atmosphere during the many long conversations I had with him.

At a more technical level, I also received much assistance from Hubert Lamb. Professor Lamb helped Ailsa Allaby and me when we were preparing the *Oxford Dictionary of Natural History* and, later, the *Oxford Dictionary of Earth Sciences*. He was unstinting with his friendly advice and I learned much from him.

I also wish to thank Frank K. Darmstadt, my editor at Facts On File. His friendly encouragement during the preparation of this *Handbook* made the work proceed very smoothly. Finally, I must thank Richard Garratt for his illustrations. The electronic wizards he keeps in his magic cave are so firmly under his control that they instantly obey his every command. At any rate, that's the way it looks from where I sit.

Michael Allaby

INTRODUCTION

All of us are interested in the weather. We need to know whether to take a coat when we go outdoors, whether we can expect a fine day for a planned outing to the beach, or whether Saturday's match will be canceled because of rain. More seriously, we need to prepare for weather that is so severe as to be dangerous. If a hurricane is approaching, if there are storms in the area so violent that they are likely to trigger tornadoes or flash floods, or if a sudden cold wave is liable to bring blizzards or ice storms, we need advance warning. For people who work outdoors, a reliable forecast can save lives. Airline pilots and sailors, for example, need to know the location and severity of storms.

Nowadays we have other reasons for being concerned about what happens in the atmosphere. Many scientists fear that by releasing carbon dioxide and other gases we may be changing the chemical composition of the air in ways that, in years to come, will affect climates all over the world.

The ongoing debate about climate change turns on the scientific understanding of atmospheric processes. Those same processes produce the everyday weather that is the subject of the nightly TV forecast—itself the product of many thousands of surface observations, satellite data, and mathematical calculations performed by some of the fastest and most powerful computers in the world. TV forecasts are not hard to follow, but sometimes they use words that sound familiar, but in fact are technical terms with precise meanings. Words like "front," "high," "low," "wind chill," and even "shower" refer to quite complicated ideas. Understanding the scientific meaning of words like these makes it easier to grasp the overall weather situation the TV presenter is describing. Similarly, scientists explaining ideas about climate change also use technical terms. Unless we know what those terms mean we will not really understand the argument. Ignorant of the underlying science, we will have no way of judging whether one person's opinion carries more or less weight than that of someone else. We will just have to accept what we are told by someone who may have a hidden reason for wanting us to take one side or the other.

The atmosphere is in constant motion, always changing. It is its ceaseless restlessness that makes weather happen, and the general type of weather that occurs year by year in a particular place constitutes the climate of that place. The scientific study of the weather is called meteorology and the scientific study of the climates of the world, in the past, present, and future, is called climatology. The two are closely related—climatologists must learn meteorology—but they are not the same.

The Facts On File Weather and Climate Handbook aims to help you understand more about the atmosphere, about the ways in which air moves, warms, cools, and gains and loses moisture. It does so by compressing a large amount of information into many short, highly succinct items. The book has four main sections.

GLOSSARY

First it is necessary to grasp the meaning of technical terms. This is the task of the Glossary, forming the first and longest part of the *Handbook.* It contains more than 2,200 entries. These are quite brief, but provide enough information to allow you to continue reading the material in which you met the puzzling word or expression.

Many of the entries are illustrated with diagrams.

BIOGRAPHIES

Like all sciences, the atmospheric sciences of meteorology and climatology have been developed over centuries by scientists who are or were real people. Some—such as Aristotle, Galileo, and Benjamin Franklin—have familiar names and the list includes some Nobel Prize winners. Others are less well known, but no less important. This section of the *Handbook* lists 100 individuals, giving the full name, nationality, years of birth and death, and achievements of each.

CHRONOLOGY

The study of the atmosphere began more than 2,000 years ago. Our word "meteorology" comes from *Meteorologica,* the title of the first book on the subject. It was written by Aristotle in about 340 BCE.

The Chronology section of the *Handbook* lists some of the most important events and discoveries in the history of atmospheric science. They are listed chronologically, starting with Aristotle's *Meteorologica* and continuing to the present day.

CHARTS & TABLES
Useful scientific information can often be summarized in a chart or table. The fourth section of the *Handbook* contains more than 20 tables of useful data. These include the Beaufort wind scale, Saffir-Simpson hurricane scale, and Fujita tornado scale; lists of the severity of avalanches and hailstorms; the albedo (reflectiveness) of different surfaces; the way the snow line changes with latitude; and many more. They also include the lists from which the names of hurricanes and cyclones in different parts of the world are taken, and the numerical code weather stations use to report the present weather conditions.

RECOMMENDED READING AND USEFUL WEBSITES
You may have turned to the *Facts On File Weather and Climate Handbook* because a book or article you were reading contained atmospheric terms that were unfamiliar to you. Perhaps that reading, and an exploration of this *Handbook,* will have sharpened your appetite for more information about the weather and climate. If so, the final section may help.

It begins with a list of books you may find useful. Many of them are textbooks of meteorology or climatology, but some are more popular books meant for background reading; they also include another Facts On File reference work: the two-volume *Encyclopedia of Weather and Climate.*

Following is a list of websites. These refer to entries in the Glossary and they are arranged alphabetically by the relevant entries (it is impossible to list Web addresses alphabetically). Like all Web addresses, these will lead you elsewhere, helping you to "surf" the vast amount of information on atmospheric science that is to be found on the Web.

I hope you will find the *Handbook* useful and fun to use.

Michael Allaby
Tighnabruaich
Scotland
www.michaelallaby.com

SECTION ONE
GLOSSARY

ablation The removal of ice and snow from the ground surface by melting and also by the process of sublimation.

absolute drought In Britain, a period of 15 consecutive days during which no rain falls.

absolute humidity The mass of water vapor present in a given volume of air, usually expressed in grams per cubic meter, but taking no account of changes in humidity caused by variations in the volume of air due to changes in temperature and pressure.

absolute instability The condition of air when the environmental lapse rate (ELR) is greater than the dry adiabatic lapse rate (DALR). As it rises, a parcel of air cools at the DALR, but because this is a slower rate of cooling than the ELR it will always be warmer than the surrounding air. If the air contains water vapor, it may reach a height at which this starts to condense to form clouds. The release of latent heat of condensation will warm the air, reducing the lapse rate from the DALR to the saturated adiabatic lapse rate (SALR). The difference between the ELR and the SALR is greater than that between the ELR and DALR, so the instability of the air will increase.

absolute momentum The sum of the momentum of a particle in relation to the surface of the Earth and its momentum due to the rotation of the Earth.

absolute stability The condition of air when the environmental lapse rate (ELR) is lower than the saturated adiabatic lapse rate (SALR). If a parcel of air is made to rise, it will cool at the dry adiabatic lapse rate (DALR). This is higher than the SALR and therefore also higher than the ELR. The rising air will quickly reach a level at which it is cooler than the surrounding air and so it will sink once more. If it is forced to rise high enough for its water vapor to start to condense, the condensation will release latent heat, warming the air and altering its lapse rate from the DALR to the SALR. The SALR is still greater than the ELR, so the rising air will always be cooler than the air around it and will sink.

absolute temperature The temperature measured on the Kelvin scale. It is reported in kelvins (K), without a degree sign (i.e., as 300K, not 300°K). The absolute temperature at which water freezes (32°F) is 273.16K and the temperature at which it boils (212°F) is 373.16K.

absolute vorticity The vorticity about a vertical axis that a mass of fluid possesses when it moves in relation to the surface of the Earth. It is the sum of the planetary vorticity (f) and relative vorticity (ζ) and in the absence of friction it remains constant, owing to the conservation of angular momentum.

absolute zero The temperature at which the kinetic energy of atoms and molecules is at a minimum. It is 0 on the Kelvin scale and equal to $-459.67°F$ $(-273.15°C)$. It is the lowest temperature possible (and unattainable according to the third law of thermodynamics).

absorption (1) A process by which one substance (the absorbent) takes up and retains another (the absorbate) to form a liquid or gaseous solution. (2) The transfer of energy from electromagnetic radiation to atoms or molecules that it strikes.

acceleration A rate of change of speed or velocity, measured in units of distance multiplied by the square of a unit time, such as feet per second per second $(ft\ s^{-2})$ or meters per second per second $(m\ s^{-2})$. If a body is moving in a straight line, and accelerating at a constant rate from a speed u to a speed v, its acceleration (a) is given by: $a = (v - u)/t$, where t is the time taken, and $a = (v^2 - u^2)/2s$, where s is the distance covered.

accessory cloud A small cloud that is seen in association with a much larger cloud belonging to one of the cloud genera. The most common accessory clouds are pileus, tuba, and velum.

acclimatization An adaptive, physiological response that allows an animal to tolerate a change in the climate of the area in which it lives.

accretion The process by which an ice crystal grows as it falls through a cloud containing many small, supercooled water droplets. If the water droplets are very supercooled, they freeze immediately on contact with the ice crystal. New crystals are added one on top of another, with air trapped between them. If the water droplets are only slightly supercooled, they may not freeze instantaneously. Instead they form a layer of liquid water that surrounds the ice crystal before freezing as clear ice.

accumulated temperature The sum of the amount by which the air temperature is above or below a particular datum level over an extended period. If, on a particular day, the mean temperature is m degrees above (or below, in which case it has a negative value) the datum level and it remains so for n hours $(= n/24$ days$)$, the accumulated temperature for that day is $mn/24$ degree days. Adding the accumulated temperatures for each day yields the accumulated temperature for a week, month, season, or year.

accumulation The extent by which the thickness of a layer of snow or ice increases over time. It represents the amount of material added, minus the amount lost during the same period through ablation.

acicular ice (fibrous ice, satin ice) Ice in the form of long, pointed crystals and hollow tubes, with air in the tubes and between the crystals.

acid deposition The placing, onto surfaces, of airborne substances that are more acid than naturally occurring, clean rain; usually as a consequence of pollution from industrial or vehicle emissions. This can occur as dry deposition, acid rain, acid mist, acid snow, or acid soot.

acidity A measure of the extent to which a substance releases hydrogen ions when dissolved in water, or the extent to which a substance acts as receptor for a pair of electrons from a base. Acidity is measured on a scale of 0–14 and is equal to $-\log_{10}c$, where c is the concentration of hydrogen ions in moles per liter. The scale measures the "potential of hydrogen," abbreviated to pH. A neutral (neither acid nor alkaline) solution has a hydrogen-ion concentration of 10^{-7} mol 1^{-1}, so it has a pH of 7. A pH lower than 7 indicates an acid solution and one higher than 7 an alkaline solution. The scale is logarithmic, so a difference of one whole number in pH values indicates a tenfold difference in acidity.

acid rain Rain that is more acidic than normal as a result of contamination by pollutants. "Acid rain" is often used as a blanket term to describe all forms of acid deposition. Ordinary, unpolluted rain has a pH of about 5.6. Acid rain has a pH value of less than 5.0.

acid soot (acid smut) Particles of soot bound together by water that has been acidified. Acid soot tends to cling to solid surfaces and is corrosive. It is a by-product of the inefficient burning of oil or coal with a high sulfur content.

actiniform An adjective describing a cloud pattern in which lines of clouds radiate from a central point or branch from one another, like the branches of a tree.

active front A weather front that is associated with appreciable amounts of cloud and precipitation.

active glacier A glacier in which the ice is flowing.

active instrument An instrument that sends out a signal that is then reflected back to it.

active layer The soil above a layer of permafrost that thaws during the summer and freezes again in winter.

actual elevation The vertical distance a weather station lies above sea level.

actual evapotranspiration (AE) The amount of water that is lost from the ground surface each month by the combined effects of evaporation and transpiration.

actual pressure The pressure measured by a barometer after it has been corrected for termperature, latitude, and any instrumental error, but before it has been reduced to the mean sea-level pressure.

adfreezing The sticking together of two objects when a layer of water freezes between them.

adiabat The rate at which a parcel of air cools as it rises and warms as it descends.

adiabatic An adjective describing a change of temperature that involves no addition or subtraction of heat from an external source.

adiabatic atmosphere A theoretical atmosphere in which the temperature decreases at the dry adiabatic lapse rate (DALR) throughout the whole of its vertical extent.

adret Sloping ground that faces in the direction of the equator and therefore is sunny.

adsorption The chemical or physical bonding of molecules of a substance (the adsorbate) to the surface of a solid object or, less commonly, of a liquid (the adsorbent), where the molecules form a layer.

advanced very high resolution radiometer (AVHRR) An instrument carried by weather satellites that senses cloud and surface temperatures. It stores its data on magnetic tape and transmits it on command to surface receiving stations. It also transmits both low- and high-resolution images in real time.

advection The transport of heat due to the movement, usually horizontal, of air or water.

advectional inversion A temperature inversion that is produced when cold air moves across a warm surface, undercutting air that had previously been warmed by contact with the surface, so that the warm air then lies above the cold air.

advection fog Fog that forms when warm, moist air is carried horizontally across a cold surface by a wind blowing at about 6–20 mph (10–32 kmh^{-1}).

advective thunderstorm A thunderstorm that is triggered by the advection of warm air across a cold surface, or of cold air above a layer of warmer air at a high level. Warm air is cooled when it crosses a cold surface, causing its water vapor to condense, releasing latent heat that makes the air unstable. Cold air moving above warm air may also cause instability by sinking beneath the warmer air and raising it, causing the warm air to cool adiabatically.

aerial plankton Bacteria, spores, and other minute organisms that are blown from the ground and carried aloft by rising air currents; they can be transported long distances.

aerodynamic roughness Irregularities in a surface that impede the passage of air and significantly reduce the wind speed up to a height equal to one to three times the height of the projecting elements causing the roughness, the magnitude of the effect decreasing with height.

aerological diagram A thermodynamic diagram on which data from soundings of the upper atmosphere are plotted. The diagram usually shows isobars, isotherms, and dry and saturated adiabats.

aerology The scientific study of the free atmosphere throughout its vertical extent, with particular reference to the chemical and physical reactions that occur at particular levels within it.

aeronomy The scientific study of the atmosphere and of the changes that occur within it as a consequence of internal or external influences.

aerosol A mixture of solid or liquid particles that are suspended in the air. Strictly speaking, a cloud is an aerosol, but the word is more usually applied to solid particles. These include soil particles, dust, salt crystals from the evaporation of sea water, smoke, aerial plankton, and organic substances. Aerosol particles are so small that they fall at about 4 inches (10 cm) a day, but they are removed much more quickly by being washed from the air by rain or snow.

Aerosol Characterization Experiments–Asia (ACE–Asia) An experiment involving more than 100 scientists, based on land and at sea, as well as satellites, that was conducted in 2001. Its purpose was to study the climatic effects of aerosols released into the air over Asia.

aerospace The whole of the atmosphere together with the adjacent region of space in which satellites orbit.

aerovane An instrument that measures wind speed and direction. The fins hold the propeller so it faces into the wind, thereby indicating the wind direction, and the speed at which the propeller spins indicates the wind speed. Both readings are converted into electrical impulses and are shown on dials.

aerovane

ageostrophic wind A wind that blows above the boundary layer at a speed different from that of the geostrophic wind predicted by the pressure gradient.

agricultural drought A drought that reduces agricultural crop yields.

agroclimatology The study of the ways in which climate affects agriculture.

agrometeorology The study of the ways in which weather systems affect farming and horticulture and the provision of weather forecasts directed at the needs of farmers and growers.

airborne dust analysis The sampling and categorization of dust particles that are seized from the air. The particles are dried and graded according to their size, either by shaking them through a series of increasingly fine sieves or by measuring their fall speeds.

aircraft ceiling (1) The height of the cloud base measured by the pilot of an aircraft flying within 1.5 nautical miles (1.725 miles, 2.78 km) of the runway of the airport to which the ceiling refers. (2) The greatest altitude to which a particular type of aircraft is able to climb.

aircraft electrification The accumulation of an electric charge on the surface of an aircraft, or the separation of a surface electric charge into charges of opposite sign on different parts of the aircraft.

aircraft thermometry The measurement of air temperature by instruments mounted on aircraft.

air drainage The downhill flow of cold air under the influence of gravity.

air frost The condition in which the air temperature is below freezing.

airglow (light of the night sky, night-sky light) A faint light that glows permanently in the night sky and is most clearly seen in middle and low latitudes. It is caused by the emission of light from molecules and atoms of oxygen, nitrogen, and sodium (from sea salt), which absorb photons from sunlight during the day. This raises them to higher energy states from which they fall back at night, emitting photons as they do so.

airlight Light that is scattered toward an observer by aerosols or air molecules lying between the observer and more distant objects. This makes the more distant objects less clearly visible. At dawn the amount of airlight increases, rendering the stars invisible. At sunset, as the amount of airlight diminishes, the stars reappear.

air mass A body of air that covers a very large area of the Earth's surface and throughout which the physical characteristics of temperature,

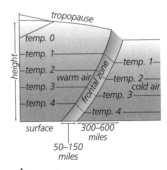

air mass

humidity, and lapse rate are approximately constant at every height. Typically, an air mass covers a substantial part of a continent or ocean and extends from the surface to the tropopause.

air mass analysis A technique in weather forecasting in which the characteristics of the air masses over a large area are related to the surface conditions shown on a synoptic chart. The aim is to build up as complete a picture of the air masses as possible in order to improve the reliability of weather forecasts based on predictions of the air masses' future behavior.

air mass climatology A type of synoptic climatology in which the weather that is typical for a region is related to the characteristics of the air masses producing it and the length of time each type of air mass remains over the region.

air mass modification The changes that take place in the characteristics of an air mass as it moves away from its source region and passes over warmer, cooler, moister, or drier surfaces.

air mass precipitation Precipitation that is wholly caused by the distribution of temperature and moisture within an air mass; it is not due to orographic lifting or the vertical movement of air in a frontal system.

air mass shower A shower that falls from a cloud, such as cumulus or cumulonimbus, which forms in unstable air.

air pocket A downdraft that causes an airplane to drop briefly but suddenly. In the early days of aviation it was supposed that the aircraft fell when it entered air that was insufficiently dense to support it. The air was likened to an empty pocket.

air pollution The release into the air of gases or aerosols in amounts that may cause injury to living organisms, including humans.

air pressure The force that is exerted over a specified area by the weight of overlying air. At sea level this averages 14.7 pounds per square inch (1 kg cm^{-2}, 100kPa, or 1 bar). Air pressure varies with the temperature and density of the air, which are related to each other by the gas laws. It also decreases with height, because the distance to the top of the atmosphere decreases and so there is less overlying air.

air quality A measure of the extent to which air is polluted. Air quality is said to be high when the concentration of pollutants is low. Pollution levels are usually related to the known damage they cause to human health, vegetation, or material structures such as buildings.

air shed The geographic area that is associated with a particular source of air. The concept is used in estimating the likelihood that the area will be exposed to air pollution.

airspace The air, measured from the surface to the uppermost limit of the atmosphere, above the surface area of a territory or nation.

airstream A large-scale movement of air across a continent or an ocean that is associated with the prevailing winds in that latitude.

air thermoscope The first thermometer, which was invented in 1593 by Galileo. It consisted of a glass bulb connected to a narrow glass tube mounted vertically with its lower end immersed in colored water held in a sealed vessel. As air in the bulb expanded and contracted with changes in the temperature, it caused the water to rise or fall in the tube. The thermoscope made no allowance for changes in air pressure, which also alter the volume of air, and so the instrument was very inaccurate.

Aitken nuclei counter A laboratory device, which is used to estimate the concentration of cloud condensation nuclei in a sample of air. The sample is drawn into a chamber, kept near saturation, and made to expand rapidly. This causes it to cool and water droplets to condense onto the nuclei present in the air, some falling onto a graduated disk, where they are counted. Each droplet is assumed to represent one condensation nucleus.

Aitken nucleus An aerosol particle with a diameter smaller than 0.4μm, most being between 0.005μm and 0.1μm. The largest act as cloud condensation nuclei.

albedo A measure of the reflectivity of a surface, expressed as the proportion of the radiation falling on the surface that is reflected. If all of the radiation is reflected, the surface has an albedo of 1.0, or 100%. Radiation that is not reflected is absorbed, so the albedo of a surface strongly influences the extent to which sunshine will warm anything below the surface.

Alberta low An area of low pressure that sometimes develops on the eastern slopes of the Rocky Mountains in Alberta, Canada. Air passing over the mountains develops a cyclonic circulation and then the system moves eastward, bringing storms with heavy precipitation.

Aleutian low A semipermanent area of low pressure that is centered over the Aleutian Islands in the North Pacific, at about 50° N. It covers a large area and generates many storms that travel eastward along the polar front and tend to merge. It is present for most of the winter and

air thermoscope

moves very little. Pressure is lowest in January, when it averages 1002 mb.

alpine glow A series of colors that are sometimes seen over mountains, especially if they are covered with snow, due to the scattering of reflected light. They appear in the east at sunset and in the west at sunrise and change from yellow to orange to pink to purple at sunset and in the reverse order at sunrise.

altimeter An instrument that measures altitude. There are two types. One relates altitude to atmospheric pressure and the other measures the time taken for electromagnetic radiation to be reflected. The type most widely used in aircraft consists of an aneroid barometer located in a pitot tube and linked to a dial on the instrument panel. The instrument computes altitude by comparing the sea-level pressure with the pressure detected by its aneroid capsules, and indicates the height above sea level. A hypsometer also measures atmospheric pressure to compute height above sea level. Radar and laser altimeters measure altitude above the ground or sea surface by transmitting an electromagnetic beam and measuring the time that elapses between the transmission and receipt of the reflection.

Altithermal A period lasting from about 8,000 to 5,000 years ago during which the average temperature was up to about 5°F (2.8°C) warmer than that of today.

altocumulus (Ac) A genus of middle cloud, composed of water droplets, that is very variable in appearance, with many species and varieties. It is white, gray, or both white and gray, and made up of elements, each about 1° to 5° across, which is approximately the thickness of three fingers held at arm's length. The elements are arranged in lines or waves, but these can be so close together that their edges merge and they form a sheet of cloud.

altostratus (As) A genus of middle cloud, composed of water droplets, that appears as a fibrous or striated veil, or a uniform sheet of cloud, and is grayish or bluish in color. The Sun or Moon can sometimes be seen through it, but it can also be thick enough to obscure them totally. Its appearance is usually an indication of approaching precipitation. A sky that is overcast with altostratus is often described as "watery."

ambient An adjective that means "surrounding." The ambient temperature is the temperature of the air surrounding the place where the measurement is made and the ambient pressure is the pressure of the surrounding air.

ambient air standard A standard for the air quality in a particular place that is defined in terms of pollution levels.

amorphous cloud A cloud such as nimbostratus, which forms a flat, featureless sheet covering most or all of the sky.

amorphous snow Snow that has an irregular crystalline structure.

amplitude The vertical distance between the mean level of a wave and the bottom of a wave trough or top of a wave crest. This is equal to half the wave height, which is the vertical distance between the base of a trough and the top of the adjacent crest.

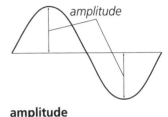

amplitude

anabaric (anallobaric) An adjective that describes any phenomenon associated with a rise in atmospheric pressure.

anabatic wind A wind that blows up the side of a hill.

ana-front A front at which the air in the warm sector is rising. Ana-fronts are very active, producing large amounts of cloud and precipitation that are often heavy.

analog model A climate model that aims to match present atmospheric conditions to times in the past when similar conditions prevailed, and then uses the way the past conditions developed to predict the development of the present conditions. Analog models are used mainly in long-range weather forecasting and in studies of global climate change.

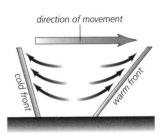

ana-front

andhis The local name for a dust storm accompanying a violent squall in the northwestern part of the Indian subcontinent.

anemogram An instrument that makes a permanent record of the wind speed. It consists of a pen linked to an anemometer and a rotating paper drum.

anemograph A record of the wind speed that is made by an anemogram.

anemology The scientific study of winds.

anemometer An instrument that is used to measure the surface wind speed. There are several types. The rotating cups anemometer is the device that is most widely used, and the aerovane is also common. A swinging-plate anemometer (pressure-plate anemometer) is preferred for measuring the speed of gusts. A bridled anemometer is often used on ships. The pressure anemometer is also used. The sonic anemometer is the type most often used at weather stations.

anemometry The scientific measurement of wind speed.

aneroid barometer

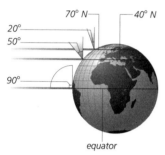

angle of incidence

angle of reflection

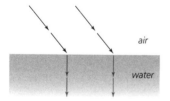

angle of refraction

anemophily The pollination of plants by the wind. Wind-pollinated plants have small flowers, usually without colored petals, and produce copious amounts of pollen. All grasses are wind pollinated. People who suffer from hay fever are allergic to pollen grains.

aneroid barometer A barometer that measures the effect of air pressure on a small, corrugated metal box from which most of the air has been removed. The box partly collapses when the air pressure increases, but a spring inside the box prevents it from collapsing completely. When the pressure decreases the box expands. The position of the surface of the box may be measured electrically, but in most aneroid barometers the surface is linked to a spring and the spring to levers that move the needle on a dial.

angle of incidence The angle at which solar radiation strikes the surface of the Earth. It is measured as the angle between the incident radiation and a tangent at the surface and it varies with latitude. At noon at each equinox the angle of incidence is 90° at the equator and elsewhere is equal to 90° minus the latitude. The angle of incidence also changes with the seasons, as the apparent position of the Sun moves north and south.

angle of indraft The angle between the direction of a steady wind and the isobars. This indicates the extent to which the wind departs from the geostrophic wind. The angle of indraft is said to be positive when the wind direction is toward the area of low pressure (which is almost invariably the case). The angle is greater over land than over sea, because friction is greater over land.

angle of reflection The angle at which light is reflected from a surface. This is always equal to the angle of incidence.

angle of refraction The angle through which light is refracted (bent) when it passes from one transparent medium to another and its speed changes. The angle varies according to the angle at which the light strikes the boundary between the two media and the difference in their refractive indices. The smaller the incident angle and the greater the difference in the refractive indices of the two media, the greater the angle of refraction.

angular momentum The momentum of a body that is moving along a curved path about an axis. Angular momentum is the product *(MΩr)* of the mass of the body *(M)*, the radius of its circle of turn *(r)*, and its angular velocity *(Ω)*, and it is a constant that is conserved (the conservation of angular momentum). If one or more of the three factors change, then one or more of the others will also change automatically, in order to conserve the constant. The consequence of the conservation of angular momentum is seen at its most dramatic in the intense speeds generated in winds spiraling toward the center of a tropical cyclone or tornado.

angular velocity (Ω) The speed of a body that is moving along a curved path, usually expressed in radians per second. The circumference of a circle is equal to 2π radians. Therefore $\Omega = 2\pi \div T$, where T is the time taken to complete one revolution. The tangential velocity (V), which is the velocity in a straight line and is measured in miles or kilometers per hour, is given by $V = \Omega r$, where r is the radius of the circle.

angular wave number (hemispheric wave number) The circumference of the Earth at a given latitude divided by the wavelength of the long waves (Rossby waves) associated with a particular weather pattern. This gives the number of waves of that wavelength that are required to encircle the Earth and, therefore, the number of times the weather pattern repeats around the world.

antarctic air Arctic and polar air masses that cover the area enclosed by the Antarctic Circle. Continental arctic air (cA) covers the continent of Antarctica, including the Antarctic Peninsula, throughout the year. In winter, maritime polar air (mP) covers the Southern (Antarctic) Ocean adjacent to the South Atlantic and Indian Oceans, and maritime arctic air (mA) covers the Ross Sea, opposite the Pacific Ocean. In summer, mP air covers the whole of the Southern Ocean.

antarctic circumpolar waves (ACW) A set of two atmospheric and oceanic waves that travel through the Southern Ocean from west to east on a track that takes them all the way around the continent of Antarctica. They move at 2.4–3.1 inches per second (6–8 cm s^{-1}) and have a period of three to five years. It takes them eight to 10 years to complete one circuit of the continent. As they pass, the waves affect wind speeds, the atmospheric pressure at sea level, the sea-surface temperature, and the location of the edge of the sea ice in latitudes south of about 25° S.

antarctic front The front that marks the boundary between arctic and polar air over the Southern Ocean. The antarctic front is almost permanent and almost continuous around the continent of Antarctica.

antarctic oscillation A periodic change that occurs naturally in the distribution of atmospheric pressure between the South Pole and latitude 55° S.

antarctic polar front (antarctic convergence, AAC) A boundary along the edge of the Southern Ocean between latitudes 50° S and 60° S where cold antarctic water sinks beneath the warmer water in higher latitudes.

antecedent precipitation index A summary of the amount of precipitation that falls each day in a particular area, weighted so it can be used to estimate soil moisture.

anthelion A spot of bright light that is occasionally seen in the sky at the same altitude as the Sun, but at the opposite azimuth. It is due to the reflection and refraction of light by hexagonal ice prisms with vertical axes.

anthropogenic An adjective that describes substances or processes produced by humans or resulting from human activities.

anticyclogenesis The stages by which an anticyclone forms or is intensified.

anticyclolysis The weakening and final disappearance of an anticyclone or ridge.

anticyclone A region in which the atmospheric pressure is higher than it is in the surrounding air. Pressure is highest at the center and decreases with distance from the center. Air flows outward from an anticyclone at a speed proportional to the pressure gradient.

anticyclonic The adjective describing the direction in which the air flows around an anticyclone or ridge. The direction is clockwise in the Northern Hemisphere and counterclockwise in the Southern Hemisphere.

anticyclonic gloom Dull, hazy conditions due to a subsidence inversion, which can develop when an anticyclone remains stationary for more than a few days. In middle latitudes, anticyclonic gloom is more common in winter than in summer, but when it does occur in summer, substances held in the inversion can react in the strong sunlight to form photochemical smog.

anticyclonic shear Horizontal wind shear that produces an anticyclonic flow in the air to one side of it.

antisolar point The position in the sky that is directly opposite to the Sun; the direction in which shadows point.

antitrade A high-level wind that blows in the tropics in a direction opposite to that of the trade winds (i.e., from the northwest and southwest in the Northern and Southern Hemispheres respectively).

antitriptic wind A wind that occurs locally or on a small scale and is caused by differences in pressure or temperature.

antitwilight arch A pink or lilac band that is seen at twilight as an arch rising to about 3° above the horizon at the antisolar point.

aphelion The point in the eccentric orbit of a planet or other body when it is farthest from the Sun. The Earth receives 7% less solar radiation at aphelion than it does at perihelion. At present, the Earth is at aphelion on about July 4, but the dates of aphelion and perihelion change over a cycle of about 21,000 years.

applied climatology The use of climatological information and concepts to help in solving economic, social, and environmental problems.

applied meteorology The preparation of weather reports and forecasts for specified groups of users, such as farmers, fishermen, aircraft pilots, climbers, backpackers, skiers, and others planning outdoor activities.

arch cloud An arch-shaped, stationary wave cloud, usually altostratus, that extends for a considerable distance along a mountain range with a wind beneath it that blows down the mountainside as a föhn wind. When seen from a distance, the arch indicates the approach of the wind.

arctic air Air that originates in the high-pressure areas of the arctic and antarctic and is very cold and dry. In winter, it is continental arctic (cA) air, with maritime arctic (mA) air off the coast of Antarctica. In summer, mA air forms over the arctic, but there is only cA air in the antarctic. Continental arctic air that forms in winter over the Arctic Basin and the Greenland ice sheet can bring cold waves characterized by extremely cold, dry, and very stable air, into North America.

arctic front (1) A boundary that exists for most of the time in northern latitudes between arctic air and more temperate air masses to the south. (2) A front that advances southward across North America and has arctic air behind it. (3) A front that forms in winter over snow and ice when the wind is weak or blows parallel to the edge of the sea ice. This front is shallow, but it can trigger the formation of a polar low.

arctic haze A reduction in horizontal visibility that sometimes occurs over the arctic. It extends to a height of more than 30,000 feet (9,150 m).

artic high A surface area of high atmospheric pressure that is centered over the Arctic Basin. The dryness of the air means there is little cloud. Strong cooling from below makes the air very stable and there is often a temperature inversion from the surface to about the 850 mb level. The high persists through the summer, but it is weaker in winter.

arctic mist　A very thin ice fog that sometimes occurs in high latitudes when the air temperature is well below freezing.

arctic oscillation (AO)　A periodic change in the distribution of pressure between the North Pole and a circle at about 55° N; when pressure is high over the Pole it is low farther south, and vice versa. The AO is said to be positive when pressure is low over the Pole. When the AO is positive, the mid-latitude westerly winds strengthen, storms travel farther north, affecting Scandinavia and Alaska, warm air is carried across Eurasia, and dry conditions predominate in the Mediterranean region and California. When the AO is negative, California and the Mediterranean region experience wet weather and the interior of Eurasia is cold.

arctic sea smoke　Fog that forms when very cold air that has crossed ice sheets and glaciers moves from the land across a sea surface that is markedly warmer. Water evaporates rapidly into the air adjacent to the warm water surface and rises by convection. This carries it into the cold air, where it condenses again to produce cloud. The "smoke" is often very dense, but it usually extends to a height of no more than about 35 feet (10 m).

arcus　A supplementary feature of cumulonimbus clouds in which the lower, darkest part of the cloud is arched. It is most commonly seen in clouds that form along squall lines.

area forecast　A weather forecast that is prepared for a specified geographic area.

argon (Ar)　A colorless, odorless gas that comprises 0.93% of the atmosphere by volume. It is a noble gas and has no true compounds. Argon has atomic number 18, relative atomic mass of 39.948, and density (at sea-level pressure and 32°F, 0°C) of 0.001 ounces per cubic inch ($0.00178 \text{ g cm}^{-3}$). It melts at a temperature of –308.2°F (–189°C) and boils at –301°F (–185°C).

aridity　Dryness; the extent to which a climate is incapable of supporting plant growth owing to the lack of precipitation.

aridity index　A measure of the extent to which the amount of water available to plants falls short of the amount needed for healthy growth. It is calculated as $100 W_d/PE$, where W_d is the water deficit and PE is the potential evapotranspiration.

aspect The direction that sloping ground faces. This determines its exposure to direct sunlight.

aspirated psychrometer A psychrometer in which the two thermometers are placed inside a tube through which air is blown.

astronomical twilight The dim daylight that illuminates areas inside the Arctic and Antarctic Circles during the early and late part of the winter. The Sun is below the horizon, but when it is less than 18° below the horizon the scattering and refraction of light allow some sunlight to reach the surface.

Atlantic conveyor A system of ocean currents that conveys cold water away from the edge of the Arctic Circle in the North Atlantic and carries warm water from the Pacific through the Indian Ocean and into the Atlantic. The conveyor is driven by the formation of the North Atlantic deep water (NADW). This sinks to the floor of the North Atlantic and flows south, across the equator and to the edge of the Antarctic Circle, where it joins the Antarctic Circumpolar Current. Part of the current turns north into the Indian Ocean and turns south again to the south of Sri Lanka, rejoining the main current. South of New Zealand the conveyor diverges from the antarctic circumpolar current, turning northward into the Pacific Ocean and rising to become an intermediate current, flowing about 3,500 feet (1,070 m) below the surface. It crosses the equator, makes a clockwise loop in the North Pacific, then travels westward through the islands of Indonesia where it crosses the equator once more, across the Indian Ocean, around Africa, and then northward, crossing the equator for the fourth time and returning to the North Atlantic. This circulation is of major importance in regulating the climates of the world.

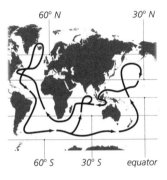

Atlantic conveyor

Atlantic highs Two anticyclones that cover a large part of the subtropical North and South Atlantic Oceans.

Atlantic multidecadal oscillation (AMO) A change in the climate over the North Atlantic Ocean that occurs over a cycle of about 50–70 years and has been detected over several centuries. Early in the 20th century it took the climate from unusually cold to unusually warm and back again. The temperature changes are of several tenths of a degree Celsius to either side of the mean. It is thought that the AMO is linked to the North Atlantic oscillation and arctic oscillation, but in ways that are not understood.

atmometer An instrument that measures evaporation. It consists of a calibrated glass tube with one end that is open to allow water to evaporate.

atmophile An adjective that describes one of the chemical elements that are concentrated in the atmosphere and that together typify its composition.

atmospheric cell A parcel of air inside which the air is moving. In the circulation of the atmosphere, Hadley cells, Ferrel cells, and polar cells are atmospheric cells.

atmospheric chemistry The scientific study of the chemical composition of the atmosphere and of the chemical processes that occur within it. Atmospheric chemists are especially concerned with the fate of pollutants released into the air and with the chemical behavior of gases that absorb radiation.

atmospheric composition The proportions of the principal gases are fairly constant throughout the lower layers of the atmosphere that comprise the homosphere. In all there are 18 gases. Nitrogen, oxygen, and argon together comprise 99.96% of the air, nitrogen being the most abundant. Water vapor and ozone are present in such widely variable amounts that proportions cannot be given. For the minor constituents, the amounts present are given in parts per million by volume (p.p.m.v.) and for the trace constituents in parts per billion by volume (p.p.b.v.). To compare these units of measurement, 1 p.p.m. = 0.0001% and 1 p.p.b. = 0.0000001%.

atmospheric dispersion The dilution of pollutants as they mix with a much larger volume of air. The rate varies according to local atmospheric conditions. Pollution occurs when atmospheric dispersion fails to reduce pollutant concentrations to levels that are harmless.

atmospheric shell (atmospheric layer) One of the layers of the atmosphere.

atmospheric structure The distinct layers that comprise the atmosphere. About half of the mass of the atmosphere is below about 3.5 miles (5.5 km) and 90% of the mass lies between the surface and a height of about 10 miles (16 km). Above 10 miles, the remaining one-tenth of the atmospheric mass extends to a height of at least 350 miles (550 km). Beyond that height the atmosphere merges imperceptibly with the atoms and molecules of interplanetary space, and especially with the outer fringes of the Sun's atmosphere, with no precisely defined upper boundary. The layers are defined by the way temperature changes with height. Above the mesopause lies the thermosphere, bounded by the thermopause at between 310 and 620 miles (500–1,000 km). Beyond the thermopause lies the exosphere.

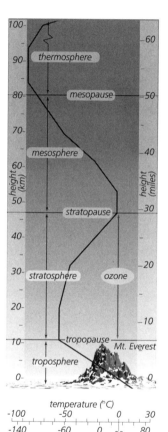

atmospheric structure

atmospheric tides Tides that occur in the atmosphere and are produced by the gravitational attraction of the Moon and Sun. The maximum value is at the equator, where the tidal variation never exceeds 0.4 lb. ft.$^{-2}$ (20 Pa). It can be measured as a twice-daily variation in atmospheric pressure, averaged over a long period, that coincides with the lunar orbit. Due to the effects of heating, the Sun exerts a much stronger effect, however, producing a maximum pressure of up to 6.3 lb. ft.$^{-2}$ (300 Pa) every morning and evening between about 9 A.M. and 10 A.M. local time.

atmospheric wave A vertical displacement in stable air in the planetary boundary layer at a height of 500–650 feet (150–200 m). Most atmospheric waves are gravity waves. Others are caused by wind shear. The height of the waves is restricted by the stability of the air, which acts to restore the displaced air to its original level.

atmospheric window The 8.5–13.0 μm waveband within which no atmospheric gas absorbs radiation. Outgoing infrared radiation from the atmosphere and surface of the Earth that has wavelengths within these limits escapes into space.

attenuation The weakening of a signal with increasing distance from its source. It is caused by the absorption of part of its energy by the medium through which the signal travels and by divergence of the signal, which spreads its energy over an increasing area.

aureole (1) A bright white or pale blue disk, surrounded by a brown ring, which is seen around the Sun or Moon. (2) A white area with no clearly defined boundary that surrounds the Sun in a clear sky.

aurora (northern lights, southern lights) Lights that are sometimes seen high in the sky, occasionally in latitudes as low as 40° but most often within oval-shaped areas in both hemispheres defined by latitude 67° at midnight and about latitude 76° at midday. They may resemble curtains hanging vertically or appear as bands, patches, or arcs of light. Auroras are caused by the interaction of solar-wind particles with atoms of oxygen and nitrogen in the upper atmosphere.

austru A cold, westerly wind that blows in winter across the low-lying plains on either side of the Danube, mainly in northern Serbia. The wind brings dry, clear air.

autan A southeasterly wind of the sirocco type that blows along the valley of the Garonne River, in southwestern France, bringing warm, moist air and rain, although it varies in strength and in the weather associated with it. Other winds in this part of southern France are sometimes called "autan," although they are not related to the true autan wind.

automated surface observing system automatic picture transmission (ASOS) A network of stations across the United States operated by the National Weather Service and comprising instruments that automatically measure surface pressure, temperature, wind speed and direction, runway visibility, cloud ceiling heights, cloud types, and precipitation intensity. ASOS units are designed to provide meteorological information for airports. They are augmented by a separate system, to provide warning of thunderstorms, run in collaboration with the Federal Aviation Administration.

automatic weather station An unmanned weather station that transmits instrument readings to a receiving center at predetermined times.

avalanche wind The wind caused by air that is pushed ahead of the descending snow of an avalanche. A major slab avalanche can generate a wind of up to 185 mph (300 kmh^{-1}). This can cause serious structural damage to buildings, so that they are already weakened when the snow reaches them.

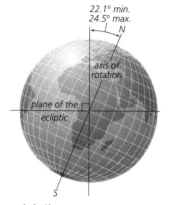

axial tilt

axial tilt The angle between the Earth's axis of rotation and a line passing through the center of the Earth that is at right angles to the plane of the ecliptic. The Earth's axial tilt varies over a cycle of about 41,000 years from a minimum of 22.1° to a maximum of 24.5°. At present the tilt is 23.45°. The axial tilt determines the location of the Tropics, which are at 23.45° N and 23.45° S, and the height of the Sun above the horizon at the summer solstice over the Poles, which is also 23.45°.

azimuth The angle between two vertical planes, one of which contains a celestial body or satellite and the other the meridian on which an observer is located. It is commonly used for reporting the position of satellites and is equal to the number of degrees from north (0°) counting in a clockwise direction.

Azores high An anticyclone that is centered over the Azores, about 800 miles (1,290 km) to the west of Portugal, often extending westward as far as Bermuda, where it is known in North America as the Bermuda high. The difference in pressure between the Azores high and the Icelandic low drives weather systems from west to east across the Atlantic. Periodic variations in this difference are known as the North Atlantic oscillation.

back-bent occlusion (bent-back occlusion) An occlusion that has reversed its direction of motion because a new cyclone has formed or the old cyclone has been shifted.

back-bent warm front (bent-back warm front) A warm front that curves around the cyclone.

backdoor cold front A cold front that brings cold air southward along the Atlantic coast toward the south and southwest of the United States. The front develops where the sea-surface temperature is much lower than the air temperature over land, and arrives from the northeast. Similar fronts travel northward along the eastern coasts of continents in the Southern Hemisphere.

backing A counterclockwise change in the wind direction. If the wind direction is given as the number of degrees from north, a backing wind decreases the number.

backscatter The proportion of a signal that is scattered from the surface at which the signal is directed. The amount of backscatter can be calculated by comparing the strength of the transmitted signal with the strength of the signal reflected from the surface to the receiver, after making allowance for attenuation and absorption by the medium through which the signal travels.

baguio The local name for a tropical cyclone in the vicinity of the Philippines or Indonesia. Baguio is the name of a town on the Philippine island of Luzon.

bai A yellow mist that occurs in parts of China and Japan. It forms when water vapor condenses onto particles of wind-blown silt, called loess. Loess is usually yellow.

Bai-u The period of heaviest rainfall during the early part of the summer monsoon season in Japan, lasting from the middle of June until the middle of July; the name Bai-u means "plum rains."

balloon ceiling The cloud ceiling when it has been determined by timing the ascent and disappearance of a balloon.

balloon drag A small balloon containing ballast that is used to retard the first part of the ascent of a larger balloon in order to allow more time for making measurements. The drag balloon bursts at a predetermined height.

balloon sounding Any measurement of atmospheric conditions made by a radiosonde.

banded precipitation (rainbands) Precipitation that varies in intensity in a pattern of bands ahead of a warm front, probably due to local instabilities along the front.

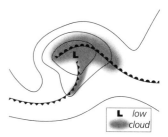

back-bent warm front

knots

———————	1–2
———————	3–7
———————	8–12
———————	13–17
———————	18–22
———————	23–27
———————	28–32
———————	33–37
———————	38–42
———————	43–47

barb

banner cloud A wave cloud that extends downwind from a mountain peak, like a flag flying from the summit.

bar A unit of pressure equal to 10^5 newtons per m^2 (= 10^6 dynes cm^{-2}). Scientists now measure atmospheric pressure in pascals (1 Pa = 1 N m^{-2}; 1 bar = 0.1 MPa), but weather reports and forecasts published in newspapers and broadcast on radio and TV still use the millibar (1 bar = 1,000 mb; 1 mb = 100 Pa).

barat A strong westerly wind on the northern coast of Sulawesi, Indonesia, during the north monsoon.

barb A line drawn at an angle at the end of a longer line to indicate a wind speed on a station model. Barbs are used only for wind speeds up to 47 knots (54 mph, 87 km h^{-1}). Pennants are used to indicate higher speeds.

barinés Westerly winds that occur in eastern Venezuela, blowing from the direction of the state of Barinas.

baroclinic An adjective describing the very common atmospheric condition in which surfaces of constant pressure and constant air density intersect. The result is that the air density changes along each isobar.

baroclinic disturbance (baroclinic wave) A change in the flow of air that results when a barotropic flow is diverted. Air then begins to flow across the isotherms, transporting cold air into a region of warmer air and warm air into a region of cooler air, producing a wave pattern in the flow of air.

baroclinic field A distribution of air pressure and the mass of a given volume of air such that the air density is not a function only of atmospheric pressure.

baroclinic instability Instability in the middle and upper troposphere associated with the waves of a frontal cyclone. If the change of wind speed with height exceeds about 3 feet per second per 3,000 feet (1 m s^{-1} km^{-1}), waves develop in the front with wavelengths that increase as the wind shear increases. Wind shear is linked to the temperature gradient, so this type of instability is said to be baroclinic.

baroclinicity The state of being baroclinic.

barogram A chart of changing atmospheric pressure produced by a barograph.

barograph (aneroidograph) An instrument that makes a continuous record of atmospheric pressure. It consists of a stack of partially evacuated boxes, of the type used in aneroid barometers, linked to a pen that inscribes a trace on a paper chart wrapped around a rotating drum.

barometer An instrument used to measure air pressure. There are four general types of barometer: the mercurial barometer, aneroid barometer, hypsometer (hypsometric barometer), and piezoresistive barometer. All barometers measure the pressure exerted by the weight of the atmosphere on a surface of unit area (square inch or square centimeter).

barometer elevation The vertical distance between sea level and the zero level of the mercury in the reservoir of a mercury barometer.

barometric law The air pressure balances the weight of all the air above the area being considered.

barometric tendency (air pressure tendency, pressure tendency) How atmospheric pressure has changed at a weather station since the last time it was reported, shown on a station model by a standard symbol to the right of the station circle.

barothermograph An instrument that records air pressure and temperature simultaneously as a pen line on a chart fastened to a rotating drum.

barothermohygrograph An instrument that records air pressure, relative humidity, and temperature simultaneously as a pen line on a chart fastened to a rotating drum.

barotropic An adjective describing the condition in which surfaces of constant atmospheric pressure and constant air density are approximately parallel at all heights. Horizontal temperature gradients are low, there is little or no change in wind direction or speed with height, and atmospheric conditions tend to be uniform over large areas.

barotropic disturbance (barotropic wave) An alteration of a barotropic atmosphere caused by a horizontal wind shear. Part of the kinetic energy in the wind shear drives an atmospheric wave.

barotropic field A distribution of air pressure and the mass of a given volume of air such that the air density is a function of only atmospheric pressure.

barotropic model A model used in numerical forecasting in which the atmosphere is assumed to be barotropic at each level. The winds are geostrophic and there is no convergence or divergence.

bath plug vortex A familiar metaphor used to describe an atmospheric vortex. Water leaving a bathtub usually forms a vortex. In it the

∧ *rising then falling*

⌐ *rising then steady or rising more slowly*

╱ *rising steadily or unsteadily*

√ *falling or steady then rising or rising then rising faster*

— *steady*

∨ *falling then rising but still same or lower than before*

╲ *falling then steady or falling more slowly*

╲ *falling steadily or unsteadily*

∧ *steady or rising then falling or falling then falling faster*

barometric tendency

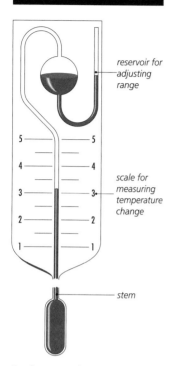

reservoir for adjusting range

scale for measuring temperature change

stem

Beckmann thermometer

conservation of angular momentum can be seen by the acceleration of the water as it nears the center and the pressure surface is drawn down the center of the vortex, just as the tropopause is drawn down to the surface in the eye of a tropical cyclone.

Beaufort wind scale A classification of winds according to their speed and effects that was devised in 1805 by Francis Beaufort, a British naval officer, to instruct the commanders of warships as to the amount of sail their ships should carry in winds of different strengths. In 1838 the Beaufort scale was introduced throughout the Royal Navy and commanders were required to record wind conditions in their daily logs.

Beckmann thermometer A mercury-in-glass thermometer with two bulbs that is used for measuring very small changes in temperature. The scale covers only about 5°C (9°F).

Beer's law When light passes through a medium, the amount that is absorbed and scattered varies according to the composition of the medium and the length of the path the light travels. This holds fairly well for the absorption and scattering of light traveling through the atmosphere and the depth to which light penetrates water, snow, and ice.

belat A strong, dusty northeasterly wind that blows across Yemen, in the south of the Arabian Peninsula.

Bergeron–Findeisen mechanism A theory of how cloud droplets grow into raindrops in clouds containing both ice crystals and supercooled water droplets. Water evaporates from the droplets and accumulates on the crystals, which grow, collide with one another, and form snowflakes. As the snowflakes fall they collide with more supercooled droplets and keep growing. Snowflakes that remain intact fall from the base of the cloud. If the temperature beneath the cloud is above freezing, the snowflakes will start to melt and some or all of them will reach the ground as rain.

berg wind A hot, dusty wind that blows across southern Africa, most frequently in winter, carrying continental air to the coast.

Bernoulli effect The reduction in air pressure when a wind blows across a convex surface, provided the flow is laminar. It follows from the Bernoulli principle, which states that the pressure within a fluid changes inversely with the speed of flow. The relationship can be expressed by $p + 1/2\rho V^2 = $ a constant, where p is the pressure, ρ is the density of the fluid, and V is the velocity. When air flows over a convex surface, it must travel farther than adjacent air that does not

force acting upward (lift)

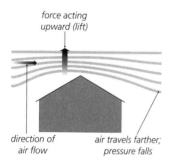

direction of air flow

air travels farther; pressure falls

Bernoulli effect

flow over the surface, but it must do so in the same time. Therefore, it must accelerate. The air must also possess the same amount of energy when it has passed the obstruction as it had before encountering it. Since its velocity increases as it passes the convex surface, the only other form of internal energy it possesses, its pressure, must decrease. In the equation above, if V increases, p must decrease.

Bilham screen A small container with louvered sides like a Stevenson screen that contains a wet-bulb and dry-bulb thermometer mounted vertically and a maximum and minimum thermometer mounted horizontally.

billow cloud A type of undulatus that consists of parallel rolls of cloud forming cloud bars separated by clear sky.

bioclimatology The scientific study of the relationship between living organisms and the climates in which they live.

biometeorology The scientific study of the relationship between living organisms and the air around them.

bise The Swiss name for a cold, dry, and fierce northeasterly wind that blows in winter and spring through the European Alps. It is often strongest in southwestern Switzerland.

Bishop's ring A faint, reddish-brown corona around the Sun caused by the diffraction of light by dust particles, usually resulting from a violent volcanic eruption.

black blizzard A dust storm in which the dust consists mainly of dark-colored soil particles.

blackbody Any object (or body) that absorbs all of the radiant energy to which it is exposed and then radiates its acquired energy at the maximum rate possible for the temperature it has reached. The energy thus radiated is known as blackbody radiation.

black frost (hard frost) The type of frost that leaves plants blackened, but with no ice crystals on external surfaces. It occurs when the air is very dry, so the temperature can fall far below freezing without causing saturation. No frost forms on exposed surfaces, but moisture freezes inside plant tissues.

black ice A layer of ice that forms when rain that is close to freezing falls onto solid surfaces that are below freezing. The water droplets spread on impact, forming a thin layer that then freezes, resulting in a fairly even covering of ice that is dark in color, hence the name.

black smoke Smoke produced when hydrocarbons are cracked (decomposed by heat) and then cooled suddenly. This releases particles of carbon, which are black.

blizzard A wind accompanied by heavy snow and a low air temperature. The National Weather Service defines a blizzard as a wind of at least 35 mph (56 kmh^{-1}), a temperature not above 20°F (–7°C), and snow that is either falling heavily enough to produce a layer at least 10 inches (250 mm) deep, or that has been blown up from the surface, and that reduces visibility to less than $1/_4$ mile (400 m). In some areas the temperature requirement has been dropped.

blob A signal on a radar screen that indicates a small-scale difference in temperature and humidity. It is produced by atmospheric turbulence.

blocking The situation in which a particular type of weather persists for much longer than usual because the movement of air that would ordinarily bring a change is obstructed or diverted.

blood rain Rain that is red because it contains red dust particles that have been transported from a distant desert region.

blowdown (windthrow) A windstorm in which trees are broken or uprooted.

bohorok A warm, dry wind of the föhn type that blows during the monsoon season along the northeastern side of the mountains of Sumatra, Indonesia.

boiling The change of phase that occurs when a liquid becomes a gas. At sea-level pressure of 1,013.25 millibars (mb), pure water boils at 212°F (100°C). At any pressure below 6.11 mb water exposed to the air cannot remain liquid, because its boiling temperature is lower than its freezing temperature, which also varies with pressure. At a pressure of 6.11 mb and a temperature of 32.018°F (0.01°C), water exists simultaneously in all three of its phases: as liquid, ice, and water vapor. This is known as the triple point for water.

bolometer An instrument that measures radiant energy by measuring the rise in temperature of a blackened metal strip placed in one of the arms of a Wheatstone resistance bridge. A bolometer can measure a temperature difference of 0.0018°F (0.0001°C).

bolster eddy (roll vortex) An eddy that sometimes forms along the foot of a steep slope or cliff on the upwind side.

bomb A cyclone that develops very rapidly over the ocean. The fronts separate and the warm front is left behind as a back-bent warm front.

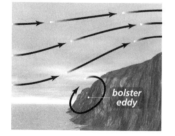

bolster eddy

bora   A cold, gusty, and usually dry north or northeasterly katabatic wind that blows along the coast of the Adriatic Sea.

bora fog A dense fog that forms when a strong bora wind lifts clouds of spray from the sea.

Bouguer's halo A faint, white arc of light, with a radius of about 39°, caused by reflection and refraction; sometimes seen at the antisolar point.

boundary layer The layer of air that lies immediately adjacent to a surface and within which atmospheric conditions are strongly influenced by the proximity of the surface.

Bowen ratio *(β)* The ratio of sensible to latent heat, which indicates how energy is apportioned at the Earth's surface. $\beta = H/LE$, where H is the sensible heat released from the surface, L is the latent heat of vaporization, and E is the energy used to evaporate water. If β is greater than 1, more energy is being released into the atmosphere as heat than is being used for evaporation.

Boyle's law The volume of a gas is inversely proportional to the amount of pressure to which it is subjected; $pV = $ a constant, where p is the pressure and V the volume. This is known as Boyle's law in English-speaking countries, but Mariotte independently reached the same conclusion, with the important addition that $pV = $ a constant only if the temperature remains constant. This improved version is known as Mariotte's law in French-speaking countries; because it is more complete, perhaps the name should also be adopted in English-speaking countries.

brave west winds The name given by sailors to the strong prevailing westerly winds that blow over the oceans between latitudes 40° N and 65° N and between 35° S and 65° S.

break   A sudden change in the weather.

breakdown potential The potential gradient of the vertical electric field in the atmosphere at which lightning will occur in dry air. It is 3×10^6 Vm^{-1}. This is 10 times greater than the largest field observed in cumulonimbus clouds, showing that it is necessary for local processes to build charge between cloud droplets and ice crystals and initiate a flash leader to begin the discharge.

breeze A light wind; on the Beaufort wind scale, a wind blowing at 4–31 mph (6.4–49.8 km h^{-1}).

brickfielder A hot, dry, dust-laden northerly wind that blows from the desert interior across southern Australia. The name "brick fielder" originated in Sydney, where winds carried dust from the brickfields to the south. When gold mines opened in Victoria, miners recruited from Sydney gave the same name to the winds they experienced there, although these blow from the opposite direction and cross no brickfields.

bridled anemometer An anemometer resembling a rotating cups anemometer, but with more cups—commonly 32. The vertical axis is bridled (i.e., its movement is checked by a spring). The rotation of the axis exerts tension on the spring and this is shown as wind speed on a dial. This type of anemometer is often used on ships.

bridled pressure plate A pressure-plate anemometer in which the movement of the plate is bridled (i.e., restrained by a spring).

bright band An enhanced radar echo from a cloud in which snowflakes are melting into raindrops.

brightness temperature A unit, with values in kelvins, used to color-code microwave satellite images of snow- and ice-covered areas. At a wavelength of 1.55 cm, for example, ice has a brightness temperature of 190K or more, but water has a brightness temperature of less than 160K. The boundary between water and sea ice shows clearly on the resulting color-coded image. Brightness temperatures vary with the wavelength of the microwave radiation. By comparing brightness temperatures at two different wavelengths, it is possible to calculate the depth of the snow covering an area.

broken The condition when 60–90% of the sky is covered by cloud.

brown smoke Smoke containing particles of tar-like compounds, produced when coal is burned at a low temperature.

brown snow Snow that is mixed with dust.

Brückner cycle A cyclical change in the weather that occurs over a period of about 35 years. Each cycle consists of a cool, moist half and a warm, dry half, with the temperature varying by not more than 2°F (1.1°C) and the rainfall by 8–9%.

Brunt–Väisälä frequency The oscillation frequency of an atmospheric gravity wave, given by $N/2\pi$, where $N = [(g/\theta)/(\partial\theta/\partial z)]^{1/2}$, g is the gravitational acceleration, θ is the constant potential temperature of the parcel of air, and $\partial\theta/\partial z$ is the vertical gradient of potential temperature.

Buchan spells Periods in the year, lasting a few days or even a week or two, when the usual rise or fall of temperature with the seasons is halted or reversed and the weather is said to be "unseasonal."

budget year The period of one year that commences with the start of snow accumulation at the firn line of a glacier. A comparison of measurements taken at the start of successive budget years indicates the growth or diminution of the glacier.

Budyko classification A climate classification based on the net radiation available for the evaporation of water from a wet surface (R_o) and the heat that would be required to evaporate the whole of the mean annual precipitation (Lr), where r is the latent heat of vaporization. The ratio of these values is used to designate climate types. The drier the climate the larger is the ratio, and unity (a ratio of 1.0) marks the boundary between dry and moist climates.

bulk modulus The ratio of the pressure applied to a fluid and the extent to which its volume decreases.

buoyancy The upward force that is exerted on a body when it is immersed in a fluid. Buoyancy occurs in air when a parcel of air has a density different from the air surrounding it. This can be expressed as $F/M = g[(\rho' - \rho)/\rho]$, where F is the buoyancy force, M is the mass of the air parcel, g is gravitational acceleration, ρ' is the density of the surrounding air, and ρ is the density of the parcel of air. Dividing F by M gives the buoyancy force per unit of mass and $(\rho' - \rho)/\rho$ is the buoyancy, often designated by B. The force exerted by the buoyancy is therefore the buoyancy (B) multiplied by the gravitational acceleration (g), or gB.

buran A fierce blizzard that occurs on the open plains of southern Russia and throughout Siberia.

burn-off The clearance of fog, mist, or low cloud during the course of the morning, as the sunshine intensifies and the temperature rises. Suspended water droplets evaporate as the dewpoint temperature rises.

burst of monsoon The abrupt onset of the summer monsoon; cool, dry weather gives way in a matter of a few hours to warm, humid air and heavy rain.

butterfly effect A metaphor that illustrates what is known formally as "sensitive dependence on initial conditions." This means that differences so small as to be undetectable can cause apparently similar weather systems to develop in radically different ways, making long-range prediction impossible. This is summarized as the suggestion that the flapping of a butterfly's wings in Brazil might set off a tornado in Texas.

Buys Ballot's law The rule that in the Northern Hemisphere, if you stand with your back to the wind there is an area of low pressure on your left. In the Southern Hemisphere, if you stand with your back to the wind the area of low pressure is on your right.

cacimbo A heavy mist or wet fog, with low stratus clouds and sometimes drizzle, that occurs along the coast of Angola during the dry season.

California fog Advection fog that affects the coastal regions of California and drifts through the Golden Gate, San Francisco, nearly every afternoon between May and October. It is driven by a sea breeze that draws warm, moist air across the cold California current.

calina A haze, due to wind-borne dust, that occurs in summer in Spain and along parts of the Mediterranean coast. Shimmer caused by the intense heat reduces visibility still further.

calm The condition in which the wind speed is 1 mph (1.6 kmh^{-1}) or less; force 0 in the Beaufort wind scale.

calm belt One of the latitudinal belts extending around the Earth where winds are usually weak or the air is calm. These belts occur in the horse latitudes, close to the Tropics, and are sometimes known as the calms of Cancer and the calms of Capricorn.

calvus (cal) A species of cumulonimbus cloud that lacks or is in the process of losing the billowing, cauliflower-like structures and cirriform appendages from its upper part.

Campbell–Stokes sunshine recorder An instrument that records the number of hours of sunshine each day. It comprises a spherical lens that acts as a burning glass, focusing the sunlight onto a card partly encircling the lens and graduated with a time-scale.

Canterbury northwester A hot, enervating northwesterly wind that blows across the Canterbury Plains of South Island, New Zealand.

cap cloud A flat-topped, cumuliform cloud that blankets a mountain peak.

Cape Hatteras low A deep depression that forms off Cape Hatteras, North Carolina, and then moves northward. It brings strong northeasterly winds and storms (nor'easters) to coastal areas from Virginia to the Maritime Provinces.

capillatus (cap) A species of cumulonimbus cloud in which the uppermost part has a fibrous or striated, cirriform structure.

capping inversion An inversion that develops when dry air advances against moist air more slowly at ground level than above the boundary layer.

The dry air overruns the moist air, preventing the development of convective clouds. Capping inversions are often associated with dry lines.

carbon cycle The cyclical movement of the element carbon through the atmosphere, living organisms, soil, rocks, and water.

carbon dioxide (CO_2) A gas formed by the complete oxidation of carbon. It is a minor constituent of the atmosphere, at present comprising 365 parts per million by volume (p.p.m.v.), or 0.0365%. It is the most important greenhouse gas, and climatic changes in the past have been associated with changes in the atmospheric concentration of CO_2.

carbon monoxide (CO) A gas formed by the partial oxidation of carbon that is emitted by volcanoes, forest fires, and the incomplete combustion of fossil fuels. CO oxidizes to CO_2 and dissolves in the oceans. It is also utilized by soil microorganisms; this is believed to be the way most of it is removed from the atmosphere.

carbon tetrachloride (tetrachloromethane, CCl_4) A clear, volatile liquid that was once widely used as a solvent, especially in dry cleaning, and also in fire extinguishers and several industrial processes. It is toxic to humans, has a global warming potential of about 1,550, and contributes to the depletion of stratospheric ozone. Its use has been banned since 1996.

carcenet (caracenet) A strong, cold wind that blows through mountain gorges in the eastern Pyrenees Mountains, on the border of France and Spain.

cardinal temperature The temperature below which almost no plant growth occurs.

castellanus (cas) A species of clouds comprising vertical protuberances, often shaped like the turrets of a castle, arising from the main cloud. These are most often seen on altocumulus, but also occur on cirrus, cirrocumulus, and stratocumulus.

ceiling The height of the lowest layer of clouds or of anything else obscuring the sky.

ceiling classification A description of the way the ceiling was determined. It is included in airport weather reports as a letter preceding the ceiling height. For example, A25 would mean an airplane had measured the height of the ceiling at 2,500 feet (762.5 m).

ceiling light A small searchlight that projects a narrow beam of light vertically upward onto a cloud base at night or onto a very dark cloud during

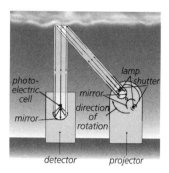

ceilometer

the day. The light illuminates a spot on the cloud. The ceiling can be calculated from the elevation of the spot.

ceilometer A device that can be used in daylight to measure the ceiling. It consists of a projector and detector. The projector has two lamps that each emit a focused beam through a shutter. The focusing mirrors and lamps rotate, so the beams are transmitted as pulses, shining at an angle onto the base of the cloud. The detector responds electronically to a series of pulses at the predetermined frequency. The height of the cloud base is calculated trigonometrically from the angles of the transmitted and reflected beams and the known distance between the projector and the detector. A ceilometer can measure cloud bases up to 10,000 feet (3,000 m) during the day and up to about 20,000 feet (6,000 m) at night.

celerity The velocity with which a wave advances. Celerity *(c)* is proportional to wavelength *(λ)* and frequency *(f)*, such that $c = \lambda f$.

Celsius temperature scale (centigrade temperature scale, °C) The scale used to measure temperature throughout most of the world, in which water freezes at 0°C and boils at 100°C. 1°C = 1K = 1.8°F.

center of action A center of high or low air pressure that is more or less permanently located in a particular place.

Central England climate record A continuous record of the mean monthly temperatures in central England since the year 1659 and a continuous record of daily temperatures in the same area since 1772. The area covered forms a triangle approximately with the cities Preston, Bristol, and London at its corners. This is the longest continuous climate record, based on instrument readings, that exists anywhere in the world.

central pressure The air pressure at the center of an anticyclone or cyclone at a particular time.

centripetal acceleration The motion of a body that is following a curved path. Although the speed of the body may remain constant, its direction constantly changes. Therefore, it is accelerating, because acceleration is defined as the rate of change of velocity, and velocity is a vector quantity that comprises both speed and direction.

CFCs (chlorofluorocarbons) A range of nonflammable, nontoxic chemical compounds in which chlorine and fluorine are bonded to carbon. CFCs were used as refrigerants, propellants in aerosol cans, solvents, and foaming agents in plastics. Their chemical stability allows them to remain in the air long enough to penetrate the stratosphere, where they are broken down by ultraviolet radiation, releasing free chlorine atoms that destroy ozone in a series of reactions. These end by releasing the original chlorine atoms and so the reactions are repeated. The production and use of CFCs is being phased out.

Chandler wobble A periodic change in the position of the Earth's axis of rotation and, therefore, of the location of the north and south geographic poles. The magnitude of the change is approximately 0.1 minutes of arc and its period is about 14 months. Its effect is to alter all latitudes by that amount.

chanduy (chandui, charduy) A cool, descending wind that blows over Guayaquil, Ecuador, during the dry season from July to November.

chaos A mathematical theory describing the behavior of dynamic systems governed by nonlinear equations of the type $y = x^2$, where a small change in the value of x produces a big change in the value of y. Consequently, the development through time of a chaotic system is acutely sensitive to very small differences in the starting conditions. This makes the behavior of the system essentially unpredictable, because those initial conditions can never be known with sufficient accuracy. If the system is observed over time, it will appear to behave randomly (i.e., chaotically). Systems that behave in this way are said to be "complex."

Charles's law The gas law stating that at constant pressure the volume of a mass of gas is directly proportional to its temperature in kelvins. This can be expressed as: $V_1/V_2 = T_1/T_2$, where V_1 and T_1 are the initial volume and temperature and V_2 and T_2 are the final volume and temperature.

chemical transport model (CTM) A three-dimensional model of the atmosphere that is used to calculate the transport of chemical substances through the atmosphere and reactions among them as a function of time.

chemisorbtion The attachment of molecules to the surface of a solid object or, less commonly, of a liquid by chemical bonds.

chergui A hot, dry wind that blows from the east across Morocco, in North Africa.

chili A hot, dry southerly wind of the sirocco type that blows over Tunisia, North Africa, most commonly in spring.

chill wind factor The wind speed in miles per hour minus the temperature in degrees Fahrenheit. For example, a wind speed of 25 mph at a temperature of –40°F would give a chill wind factor of: 25 – –40 = 25 + 40 = 65.

chimney plume The cloud of gases, water droplets, and solid particles that is emitted from a chimney or factory smokestack and travels downwind.

chinook A warm, dry, katabatic wind that blows on the eastern side of the Rocky Mountains in Canada and the United States, mainly in late winter and spring.

chromosphere The gaseous layer of the Sun (or any other star) that lies above the photosphere. The temperature rises through the solar chromosphere from about 4,000K at the base to about 10,000K at the top.

chubasco A heavy thunderstorm with strong squalls that occurs along the western coast of Central America, especially in Nicaragua and Costa Rica, between May and October.

circular vortex A vortex in which the streamlines are parallel to one another.

circulation The horizontal or vertical movement of air or water along a path that eventually returns it to its starting point. Air circulates vertically in a convection cell and the general circulation of the atmosphere involves a number of large-scale vertical cells that are combined in the three-cell model of the general circulation. The horizontal circulation of air may be cyclonic or anticyclonic and the large-scale circulation may consist mainly of zonal or meridional flow.

circulation flux Flux associated with the overall movement of the atmosphere.

circulation index A value ascribed to one of the major components of horizontal circulation. There are two such indices, the meridional index and the zonal index. Zonal flow also varies over an index cycle.

circulation pattern The geometric shape of the horizontal circulation of the atmosphere as this is shown by the isobars on synoptic charts.

circumhorizontal arc A horizontal band of light, displaying the colors of the spectrum with red at the top, seen at an elevation of less than 32° above the horizon when the Sun is a little more than 58° above the horizon. It is caused by the reflection and refraction of light from ice crystals with vertical axes.

circumpolar vortex The circulation pattern formed by the westerly winds around the North and South Poles. The winds circulate cyclonically around a persistent region of low pressure at an altitude of 6,500 to 33,000 feet (2–10 km).

circumscribed halo A halo surrounded by a bright ring. It is caused by the refraction of light through hexagonal ice crystals with horizontal axes.

circumzenithed arc A circular arc, brightly colored with red at the bottom, seen more than 58° above the Sun when the Sun is below 32°. It is caused by light rays entering the horizontal tops of hexagonal ice crystals with vertical axes and emerging from vertical sides.

cirriform Stretched into long, fine, curling filaments that resemble cirrus clouds.

cirrocumulus (Cc) A genus of high clouds composed entirely of ice crystals that appears as small, white patches or sheets, or as more or less spherical masses, called elements, each with an apparent width of about 1°. The elements are arranged in more or less regular patterns resembling the ripples seen in sand on the seashore or, less commonly, form groups or lines.

cirrostratus (Cs) A genus of high clouds composed entirely of ice crystals that appears as a thin, white veil, which does not blur the outlines of the Sun or Moon, although it often gives rise to halos. Sometimes it is so thin that it does no more than give the sky a pale, milky appearance. At other times it has a distinctly fibrous appearance.

cirrus (Ci) A genus of high clouds composed entirely of ice crystals that appears as long, wispy filaments, narrow bands, or white patches, always with a fibrous appearance.

Clausius–Clapeyron equation An equation that relates the saturation vapor pressure (e_s) to the absolute temperature (T). The equation is: $de_s/dT = L/T(\alpha_2 - \alpha_1)$, where L is the latent heat of vaporization, α_2 is the specific volume of water vapor and α_1 is the specific volume of liquid water. Since α_2 is usually very much larger than α_1, α_1 can be ignored.

clear air turbulence (CAT) Vertical air currents that occur in unstable air that is not saturated and therefore free from cloud. It can be caused in several ways. Wind shear associated with the jet stream is the commonest cause of the clear air turbulence that occasionally affects aircraft.

climate The average weather conditions experienced in a particular place over a long period. The climate of a place is determined principally by its latitude, distance from the ocean, and elevation above sea level.

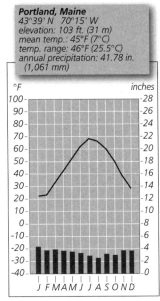

Portland, Maine
43°39' N 70°15' W
elevation: 103 ft. (31 m)
mean temp.: 45°F (7°C)
temp. range: 46°F (25.5°C)
annual precipitation: 41.78 in.
 (1,061 mm)

climate diagram

climate classification The arrangement of climates according to their most important characteristics in order to provide each type with a short, unambiguous name or title by which it can be known.

climate controls The factors that determine the type of climate a particular place will experience.

climate diagram A diagram that shows the mean temperature and precipitation month by month for a particular place. The diagram also shows the name of the place to which the data refer, often with its latitude and longitude, and additional information may also be displayed, such as height above sea level, the total annual precipitation, and temperature range.

Climate–Leaf Analysis Multivariate Program (CLIMAP) A program that uses plant leaves to estimate past temperatures, based on the strong correlation that has been observed between the warmth of the climate and the likelihood that dicotyledonous plants will have smooth-edged leaves.

Climate: Long Range Investigation, Mapping, and Predictions (CLIMAP) An international scientific project that ran from 1971 until 1980 and aimed to reconstruct the climates of the Quaternary. CLIMAP has been succeeded by the Cooperative Holocene Mapping Project.

climate model A mathematical simulation of the processes affecting the atmosphere that produce local weather and the climates of large regions, or the entire world, over extended periods. The model consists of a computer program in the form of a series of equations that allow the physical laws controlling the weather to be applied. It is used as an aid to understanding those processes and predicting how changes to them may affect weather and climate.

climate system The atmosphere, together with all the factors that affect it to produce the climates of the world.

climatic divide The boundary between two or more regions that have markedly different types of climate.

climatic elements The factors that determine the weather and climate. These include temperature, sunshine, air pressure, wind direction and speed, humidity, cloudiness, and precipitation. Visibility is also a climatic element, but of less general importance.

climatic forcing A perturbation of the balance between the amount of energy the Earth receives from the Sun and the amount it re-radiates back into space, which is imposed by some factor outside the climatic system that produces climatic effects. Climatic forcing is measured in watts per square meter.

climatic geomorphology A branch of geomorphology that concentrates on the influence of climate in shaping the surface of the planet.

climatic normal The mean values for temperature, humidity, and precipitation at a specified place over a fixed period. In many countries, including the United States, the fixed period is 30 years and the period changes every 10 years.

climatic optimum A period during which the climates over most or all of the world are warmer than those before or after. The medieval warm period was a climatic optimum, but the warmest period since the end of the most recent ice age occurred between about 7,000 and 5,000 years ago, when summer temperatures in Antarctica and in Europe were about 4–5°F (2–3°C) warmer than those of today. This postglacial optimum reached Greenland and northern North America about 4,000 years ago.

climatic prediction An estimate of the possible climatic consequences of social or industrial changes that are already occurring and that are likely to continue.

climatic region (climatic province) A large area over which the climatic normals are fairly constant, so the normals from different stations can be grouped together.

climatic snow line The altitude above which snow will accumulate over a long period on a level surface that is fully exposed to sunshine, wind, and precipitation. Below this altitude, ablation between snowfalls will be sufficient to prevent snow from accumulating.

climatic zone A region of the Earth, defined by latitude, within which the climate is sufficiently constant to be characteristic of the region as a whole.

climatological forecast A weather forecast for a region based on its climate, but with allowance for such important features as fronts, pressure systems, and the location and strength of the jet stream.

climatological intra-seasonal oscillations (CISOs) A series of weather cycles that bring alternately wet and dry conditions to regions affected by the Northern Hemisphere summer monsoon.

climatological station An observing station where meteorological data are collected and stored over a long period for use in climatological studies.

climatology The scientific study of climates. This encompasses every aspect of the physical state of the atmosphere over particular parts of the world and over extended periods of time.

climatostratigraphy The study of traces of soil and living organisms found in sedimentary rocks that were formed during the Quaternary. These rocks can be dated and the fossils and other materials found in them provide clues to the climatic conditions when the sediments were deposited.

close (oppressive, muggy, sticky, stuffy) A subjective feeling of discomfort that people sometimes experience when the air is still and warm. It is caused by a combination of high temperature and a relative humidity high enough to inhibit the evaporation of sweat from the skin.

clothesline effect Advection, when warm, dry air enters and flows through vegetation, such as a forest or farm crop. Near the edge the moving air raises the temperature and the rate of evaporation. This has a drying effect on the soil. Farther into the vegetation stand, the air cools, raising its relative humidity.

cloud A large concentration of liquid water droplets or ice crystals that form by the condensation of water vapor in saturated air and that remain suspended in the air, clear of the surface. At any time about half the surface of the Earth is covered by clouds.

cloud amount The extent to which the sky is obscured by cloud. This is reported from weather stations together with details of the cloud type. Cloud amount is measured from a reflection of the sky in a mirror divided into equal areas by grid lines. The observer counts the number of these areas that are filled with cloud.

cloud band A linear formation of clouds 10–100 miles (16–160 km) wide and tens to hundreds of miles long.

cloud bank A well-defined mass of cloud that is seen from a distance extending across most of the horizon but not covering the sky overhead.

cloud bar A clearly defined, long, and narrow horizontal cloud.

cloud base The height of the lowest part of an individual cloud or a layer of cloud, measured as the vertical distance above sea level.

cloud bow An arc of light seen in the sky, similar to a fog bow, that is caused by the refraction of light through spherical water droplets.

cloudburst A sudden, very intense shower of rain that occurs when the mechanism sustaining a cumulonimbus cloud fails and the cloud starts to dissipate, losing all of its moisture as it does so. Individual cloudbursts are usually of brief duration, but they may be repeated, because as one cloud dissipates another forms.

cloud chamber A device that is used to study cloud formation and to detect the tracks of charged particles.

cloud classification A method for arranging cloud types for purposes of identification. Clouds are classified first according to the height of the cloud base, as high, middle, or low. This division is made mainly for convenience and refers to the height at which the bases of clouds most commonly occur. Then clouds are divided according to their appearance into 10 basic types, or genera, with names based on the Latin names "cirrus," "cumulus," and "stratus." These mean "hair," "pile," and "layer" (from *stratum*), respectively.

cloud condensation nuclei (CCN) Small particles that are carried by the air and onto which water vapor condenses, the resulting droplets forming clouds.

cloud deck The upper surface of a cloud layer.

cloud discharge A flash of lightning between areas of positive and negative charge within a single cloud. From the ground a cloud discharge appears as sheet lightning.

cloud droplet A particle of liquid water held in suspension inside a cloud. A typical cloud droplet is about 0.0004 inch ($10\mu m$) in diameter, there are about 283 of them in every cubic foot of air (100,000 per liter), and they fall at about 0.4 inch per second (1 cm s^{-1}).

cloud echo A radar signal that has been reflected by cloud droplets.

cloud formation (1) The processes by which clouds form. (2) A pattern, or formation, of clouds with particular shapes.

cloud height The vertical distance between the surface and the cloud base.

cloud layer Stratiform clouds, or clouds of the same or different types, that cover all or part of the sky and have a cloud base at approximately the same height everywhere. Cloud layers can occur at different levels, one above the other, with clear air between them.

cloud level One of the three groups (high, middle, and low) into which clouds are classified by the height of the cloud base.

cloud particle A liquid droplet or ice crystal that forms part of a cloud.

cloud physics The scientific study of the physical properties and behavior of clouds.

cloud searchlight A powerful light used to measure the height of the cloud base. The light shines vertically upward, illuminating an area on the cloud. An observer some known distance away measures the elevation of the center of the illuminated spot. The height of the cloud base can then be calculated by trigonometry.

cloud seeding Injecting material into supersaturated air to make water vapor condense into cloud droplets or ice crystals. The technique is used to make rain fall where it might not have fallen otherwise and also to protect farm crops by inhibiting the formation of hail.

cloud shadow Shading of the ground by a cloud that is overhead. This is the primary cause of reduced sunshine.

cloud street A row of small cumulus clouds aligned with the wind direction, most often seen in the early morning and evening. Small surface irregularities trigger the development of thermals that are carried downwind and form a series of convection cells. When the ground surface is cool and convection weak, a breeze may cause the cells to merge into a series of convective spirals traveling downwind. If, at the top of each spiral, rising air cools to below its dewpoint temperature, water vapor will condense to form a cloud. Before the cloud has time to grow, the air has descended on the downward side of the spiral and its temperature has risen to above its dewpoint.

cloud symbols A set of ideograms used on station models to indicate the type of cloud observed.

cloud top The highest altitude at which there is a perceptible amount of a particular cloud or cloud layer.

cloudy The condition in which more than 70% of the sky (6 oktas) remains covered by cloud for at least 24 hours.

cluster map A weather map on which the situation over a large area, such as the coterminous United States, is shown for a particular day. It is prepared by a statistical technique that "clusters" data from many weather stations.

coalescence The merging of two or more cloud droplets into a single, larger droplet.

coalescence efficiency The proportion of colliding cloud droplets that merge to form larger drops. The higher the coalescence efficiency, the more rain the cloud will produce.

direction of convective airflow   cloud streets

wind direction

cloud street

cirrus	cirrostratus
cirrocumulus	altostratus
altocumulus	nimbostratus
stratocumulus	stratus
fractostratus	cumulus
towering cumulus	cumulonimbus

cloud symbols

coastal desert A desert adjacent to a coast. The Atacama and Namib Deserts, parts of the Sahara Desert, and the desert of Baja California are coastal deserts. These are the driest of all subtropical deserts. Air approaching from the ocean crosses a cold boundary current with upwelling flowing parallel to the coast. The lowest layer of air is cooled by contact with the cold water, increasing its stability, producing a shallow inversion, and often chilling it to below its dewpoint temperature. This produces frequent fog and low cloud. Although the air is moist, the clouds rarely produce rain, because the inversion inhibits convection.

coefficient of haze (COH) A measure of the air pollution caused by small particles, calculated from the proportion of the total amount of light that is transmitted through a filter paper that has been exposed to the air for a specified length of time.

coefficient of viscosity The force per unit area, applied at a tangent, that is needed to maintain a unit relative velocity between two parallel planes set a unit distance apart in a fluid.

cohesion The force that causes water molecules to become attached to each other by hydrogen bonds. Cohesion causes additional water molecules to attach themselves to the molecules coating surfaces, such as exposed mineral particles in soil.

col The area between two centers of high or low pressure where the pressure gradient is low.

cold anticyclone (cold high) An anticyclone in which the air at the center is colder than the surrounding air.

cold cloud A cloud in which the temperature is below freezing throughout.

cold core The center of a depression associated with fronts that have occluded; located on the poleward side of the warm air mass.

cold front A boundary between warm and cold air that advances with the warm air ahead of the cold air. Air behind a cold front is cooler than the air ahead of the front, but "cold" implies no particular temperature.

cold-front thunderstorm A thunderstorm produced on a cold front. As cold air pushes beneath warm air or warm air rises over cold air, moist air in the warm sector becomes sufficiently unstable to generate cumulonimbus clouds that cause storms.

cold lightning Lightning that does not ignite forest fires, unlike hot lightning. The current carried by the lightning stroke releases heat beneath the

cold front

bark of trees. This can cause local explosions that strip away sections of the bark, but the current is not sustained into the return stroke.

cold low (cold pool) A depression that consists of cold air surrounded by warmer air at a higher pressure. Cold lows often form in winter in the middle troposphere over northeastern North America and northeastern Siberia, and they are usually persistent.

cold pole One of the places that experience the lowest mean temperatures on Earth. The cold poles do not coincide with the geographic North or South Poles. In the Southern Hemisphere, the cold pole is at Vostok Station, 78.46° S and 106.87° E, where the temperature on July 21, 1983, was –128.6°F (–89.2°C); this is the lowest surface temperature that has ever been recorded on Earth. The annual mean temperature at Vostok is –67.1°F (–55.1°C). The Northern Hemisphere cold pole is at Verkhoyansk, Siberia, at 67.57° N and 133.85° E, where the mean annual temperature is 1.1°F (–17.2°C) and the lowest temperature ever recorded is –89°F (–67°C). There is also a cold pole in North America, at Snag, Yukon, in northwestern Canada, at about 62.37° N and 140.40° W. In February 1947 the temperature at Snag airport fell to –81°F (–63°C); this is the lowest temperature ever recorded in North America. The mean annual temperature at Snag is 21.6°F (–5.8°C).

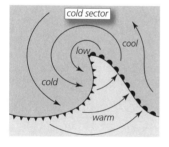

cold sector

cold sector The cold air that partly surrounds the warm sector during the development of a frontal system.

cold tongue A narrow, tongue-like extension from a cold air mass that protrudes in the direction of the equator.

cold water sphere (oceanic stratosphere) That part of the ocean where the temperature is lower than 46°F (8°C).

cold wave A sudden and large drop in temperature. Over most of the United States, a cold wave is defined as a temperature decrease of at least 20°F (11°C) that occurs over a period not exceeding 24 hours and reduces the temperature to 0°F (–18°C) or lower. In California, Florida, and the Gulf Coast states, the temperature drop must be of at least 16°F (9°C) to a temperature of 32°F (0°C) or lower.

colla (colla tempestade) A southerly or southwesterly wind over the Philippines, blowing at up to 39 mph (63 kmh⁻¹), accompanied by heavy rain and severe squalls.

collada A moderate gale from the north or northwest over the northern part of the Gulf of California and from the northeast over the southern part.

collision efficiency The proportion of cloud droplets that collide with other droplets. In order to collide, the small droplets must be very close to the center of the path followed by the large ones. The higher the collision efficiency, the more rain the cloud will produce.

collision theory The theory describing the way raindrops form in warm clouds. These contain droplets of varying sizes. The larger ones fall faster than smaller ones. As they fall, they collide and coalesce with smaller droplets in their path.

colloid Two homogeneous substances in different phases that are thoroughly mixed together. A cloud is a colloid consisting of water droplets (liquid) distributed in air (gas).

colloidal instability The property of clouds that causes water droplets to aggregate into larger drops.

Colorado low An area of low pressure that sometimes develops on the eastern slopes of the Rocky Mountains, in the state of Colorado.

combustion A chemical reaction in which an element combines rapidly with oxygen and energy is released in the form of heat, light, or both.

combustion nucleus A cloud condensation nucleus that was released by combustion.

comfort zone The range of temperatures within which humans feel comfortable. For most people this is between about 65°F (18°C) and 75°F (24°C), but wind and relative humidity can make the air feel warmer or colder than it really is. Below an apparent temperature of 80°F (27°C), there is no risk to most people. At an apparent temperature of 80–90°F (27–32°C), caution should be exercised. At an apparent temperature of 90–106°F (32–41°C), extreme caution should be exercised. An apparent temperature of 106–130°F (41–54°C) is dangerous. An apparent temperature in excess of 130°F (54°C) is extremely dangerous.

complex low A region of low air pressure with more than one low-pressure center.

compressional warming The mechanism by which a fluid warms when it is compressed, due to the transfer of kinetic energy to the compressed fluid from the surrounding fluid that exerts the compressional force.

concrete minimum temperature The lowest temperature registered by a minimum thermometer that remains in contact with a concrete surface for a specified period.

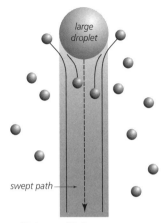

collision efficiency

condensation The change of phase in which a gas is transformed into a liquid.

condensation temperature (adiabatic condensation temperature, adiabatic saturation temperature) The temperature at which a parcel of air will reach saturation if it cools at the dry adiabatic lapse rate (DALR).

conditional instability The condition of air when the environmental lapse rate (ELR) is greater than the saturated adiabatic lapse rate (SALR), but lower than the dry adiabatic lapse rate (DALR). The air remains stable until it is forced to rise to its lifting condensation level. Water vapor then starts to condense, releasing latent heat that warms the air and slows its lapse rate from the DALR to the SALR. This is lower than the ELR, so the rising air will always be warmer than the surrounding air. Being warmer, it will remain buoyant and continue to rise. It is then unstable.

conditional instability of the second kind (CISK) The type of conditional instability that leads to the formation over tropical oceans of large, long-lived clusters of cumulonimbus clouds that sometimes develop into tropical cyclones. It is caused by the intensification of conditional instability by horizontal convergence. This triggers convection, at the same time feeding moisture into the convection cells and pushing the developing clouds closer together. Divergence above the clouds reduces the surface pressure, which intensifies and sustains the low-pressure system.

conduction The transmission of heat through a substance from a region that is relatively warm to one that is relatively cool until both are at the same temperature.

confluence A flow of air in which two or more streamlines approach one another and the air accelerates.

congestus (con) A species of cumulus clouds that is large and growing rapidly, usually by the development of towering, billowing structures in the upper parts of the cloud. A cumulus congestus (Cu_{con}) cloud looks like a cauliflower.

coning A downwind pattern made by a chimney plume. The plume widens with increasing distance from the smokestack and a line through the center of the plume is horizontal. Coning occurs when the wind is fairly strong and the plume is moving through stable air.

constant gas A constituent atmospheric gas that is present in the same proportion by volume to an altitude of about 50 miles (80 km).

coning

constant-height chart A synoptic chart on which the meteorological conditions at a particular altitude are plotted, based on radiosonde data. The heights most often used are 5,000, 10,000, and 20,000 feet (1,525, 3,050, and 6,100 m).

constant-level balloon A weather balloon that is designed to rise to a predetermined altitude and then remain there.

constant-level chart A synoptic chart on which meteorological conditions at a particular level are plotted.

constant-pressure chart A map showing the distribution of atmospheric pressure that assumes the existence of a level surface where the pressure is constant. Pressure contours indicate heights above or below this imaginary surface. The resulting chart resembles an isobaric map, but its lines represent heights, not pressures.

constant-pressure surface (isobaric surface) A surface across which the atmospheric pressure is everywhere the same.

contact cooling The cooling that occurs when warm air comes into contact with a surface at a lower temperature.

contessa del vento A lenticular cloud with a rounded base and bulging upper surface.

continental air Air that forms very dry air masses over all the continents.

continental climate A climate produced by continental air that occurs in areas deep in the interior of continents. Continental climates are dry, and hot in summer and cold in winter.

continental high An area of high pressure that covers the center of a continent.

continentality The extent to which the climate of a particular place resembles the most extreme type of continental climate. The index of continentality (K) can be calculated from $K = 1.7 \, A/\sin (\varphi + 10) - 14$, where A is the average annual daytime temperature range and φ is the latitude. An extreme continental climate has a value of 100.

continentality effect The production of extreme summer and winter temperatures in water that is almost completely surrounded by a continental land mass. This produces a continental climate in places that would otherwise experience a maritime climate owing to their proximity to an ocean.

contingent drought (accidental drought) A drought that is unpredictable, can occur anywhere, and ends no more predictably than it starts.

continuous-wave radar Radar that transmits a continuous pulse. It is used to determine the speed of a moving object.

contour microclimate A variation in a microclimate that is directly due to a difference in elevation.

contrail (condensation trail, vapor trail) A long, narrow cloud produced by the condensation of water vapor in the exhaust from an aircraft engine. They are usually composed of ice crystals. Most contrails form above 20,000 feet (6,000 m), but the precise altitude at which they form varies from day to day. Where the air traffic is heavy, contrails increase cloudiness by an appreciable amount.

contrastes Winds that blow along the shore, but from opposite directions on the two sides of a headland. They occur in winter along the Mediterranean coast of Europe.

control day A day on which folk custom holds that the weather will determine the weather over the coming weeks, months, or season. Many control days are associated with Christian saints or festivals.

convection Transport of heat through vertical motion within a fluid. A convective circulation is driven by gravity.

convectional precipitation Rain, snow, or hail showers resulting from thermal convection in moist air.

convection cell A form of circulation in which warm air rises, cools, and subsides again.

convection street A type of convection cell that is drawn out into long, parallel lines by a strong wind near the surface and wind shear at a high level.

convective cloud A cloud that develops vertically due to convection. Cumuliform clouds are of this type.

convective condensation level The height at which condensation occurs in a parcel of air that becomes saturated while rising by convection through air in which temperature falls at the dry adiabatic lapse rate (DALR) and there is conditional instability above the height at which the parcel becomes saturated.

convective equilibrium (adiabatic equilibrium) The condition in a tall column of air that is being mixed predominantly by mechanical processes and convection. Due to mixing, any parcel of air within the column is at the same temperature and pressure as the air around it. If it is displaced vertically, the parcel expands or is compressed, so it remains at the same temperature and pressure as the surrounding air.

Convective equilibrium also ensures that the molecules of atmospheric gases that remain do not separate, and solid particles remain evenly distributed.

convective inhibition The energy a parcel of air needs to overcome an inversion that inhibits its rise by convection.

convective instability (potential instability, static instability) Instability caused by convection in a stratified atmosphere where the potential temperature decreases with height and the lower part of a rising layer of moist air becomes saturated before the upper part. Rising air in the lowest layer cools at the dry adiabatic lapse rate (DALR), saturated air in the layer above it cools at the shallower saturated adiabatic lapse rate (SALR), and air in the uppermost layer cools at the DALR. If the lapse rate through the entire layer of rising air is greater than the SALR, saturated air will always be rising into air that is cooler. The entire layer will then be unstable and may overturn. As it continues to rise, however, it will also cool and eventually the atmosphere will become statically stable.

convective region An area of the surface above which air is rising by convection, or where convection is especially common.

convergence A flow of air in which streamlines approach an area from different directions. The increased pressure causes air to rise, so a region of low-level convergence is also a region of rising air. The rising air eventually reaches an inversion level where air spreads out, moving away from an upper-level area that corresponds to the area of low-level convergence.

convergence line A horizontal line along which convergence is occurring.

conveyor belt A wind that blows up the slope of a front. The warm conveyor belt carries air over the warm front and then turns until it is approximately parallel to it. The cold conveyor belt blows ahead of the warm front, carrying air from the cold sector.

cooling degree days A measure used in calculating the power needed to cool buildings. It assumes that 65°F (18°C) is a comfortable temperature and that cooling will be required whenever the temperature rises above that value. Cooling degree days are registered in degrees Fahrenheit and counted by subtracting 65°F from the mean temperature day by day. If the temperature rises to 75°F, for example, 10 cooling degree days will be recorded on that day. At the end of the year the cooling degree days for each day are added together to give an annual total.

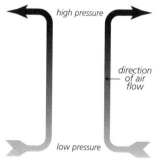

high pressure

direction of air flow

low pressure

convergence

Cooperative Holocene Mapping Project (COHMAP) A long-term study of the climatic changes that have taken place over the last 18,000 years.

cooperative observer In the United States, a voluntary weather observer who maintains a weather station and supplies data to the National Weather Service without remuneration.

cordonazo de San Francisco A hurricane that forms over the Pacific Ocean, off the coast of Central America between Costa Rica and Point Eugenio in Baja California, Mexico.

coreless winter A winter in which the temperature falls to its minimum at the autumn equinox, after which it falls no further. This happens only in Antarctica.

Coriolis effect (CorF) The deflection a body that is not attached to the surface of the Earth experiences when it moves with respect to the surface. The deflection is to the right in the Northern Hemisphere and to the left in the Southern Hemisphere, and is due to the rotation of the Earth. The magnitude of the CorF is zero at the equator and reaches a maximum at each pole. The change in CorF magnitude with latitude is given by the Coriolis parameter. This is $2\Omega \sin \varphi$, where Ω is the angular velocity of the Earth (7.29×10^{-5} rad s^{-1}) and φ is the latitude. CorF also varies according to the speed of the moving body. When this is taken into account the magnitude of the CorF is $2\Omega \sin \varphi v$, where v is the speed. The result is an acceleration, because Ω, in units of radians per second, is multiplied by v, in units of meters (or feet) per second, to give a result measured in units of distance per second per second.

corner stream Air that is deflected around the sides of a tall building that is downwind of a lower building when a wind is blowing.

corona A whitish disk surrounding the Moon or less commonly the Sun caused by the diffraction of light through the water droplets in a layer of cloud, commonly altostratus. Colors can sometimes be seen, in which case the corona consists of two or more concentric rings that are reddish on the outside and bluish on the inside.

corrected altitude (true altitude) The altitude measured by an altimeter and adjusted to take account of the difference between the ambient temperature and the temperature of the standard atmosphere.

cosmic radiation A stream of high-energy particles originating in space, some of it outside the solar system, that falls onto the Earth. Primary cosmic rays consist of nuclei of the commonest elements, predominantly of hydrogen, as well as electrons, positrons (particles identical to electrons, but carrying a positive charge), neutrinos, and photons (electromagnetic radiation) at gamma-ray wavelengths of about 0.00001 μm.

cotton-belt climate A climate with warm, wet summers and dry winters, characteristic of the cotton-growing regions of the southern United States and China.

country breeze A light, cool wind that blows into a city from the surrounding countryside, especially on calm nights when the sky is clear, due to the urban heat island effect.

crachin Fog, accompanied by low stratus cloud and drizzle, that is common between February and April along the coast of southern China and the Gulf of Tonkin.

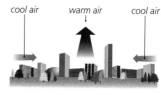

cool air   warm air   cool air

country breeze

crepuscular rays (Sun drawing water) Rays or bands of light that radiate upward from the Sun when the Sun is low and the sky partly covered by cloud, due to the illumination of particles in its path by light shining through gaps in the cloud. People used to believe the rays were caused by water being drawn from the sky toward the Sun and described the phenomenon as "the Sun drawing water," taking it as a sign of fine weather to come. This is mistaken, but crepuscular rays can be seen only when there are gaps in the cloud cover, which often means the cloud is breaking up and the weather is turning fine.

crest cloud (cloud crest) A cloud that marks the crest of a lee wave.

criador A westerly wind that brings rain in northern Spain.

critical point The temperature above which a gas cannot be liquefied, regardless of the pressure to which it is subjected.

crivetz A northeasterly wind that brings hot weather in late spring and early autumn and cold weather, sometimes with snow, during the colder months to the lower Danube region in Romania and southern Ukraine.

crop yield model A climate model that relates local weather conditions to crop yields.

crosswind A wind that is blowing neither in the same direction as a moving object, such as an airplane, nor in the opposite direction.

cryogenic limb array etalon spectrometer (CLAES) An instrument carried by the Upper Atmosphere Research Satellite that measures infrared radiation emitted from the atmosphere.

cryosphere Snow and ice that lie on the surface of continents and oceans.

cryptoclimatology The climatology of enclosed spaces.

cumuliform An adjective describing the shape of a cloud resembling cumulus or cumulonimbus.

cumulonimbus (Cb) A genus of dense cloud with a low base that often extends vertically to a great height, sometimes all the way to the tropopause. The uppermost part is usually smooth and the top flattened, marking the level beyond which air is unable to rise by convection. Cumulonimbus brings precipitation, which is often heavy, and it is associated with thunderstorms, tropical cyclones, and tornadoes. The great depth of cloud in which light is scattered by water droplets makes the lower part of a cumulonimbus cloud dark and menacing.

cumulus (Cu) A genus of fleecy or billowing cloud that develops vertically as warm air rises by convection. Blue sky is often visible between cumulus clouds, so their boundaries are sharply defined. Small, scattered cumulus clouds seen on a fine day are known as "fair weather cumulus." At other times cumulus may be immersed in clouds of other types, so it is more difficult to distinguish.

current A flow of fluid (gas or liquid) through a mass of a similar fluid that remains stationary. Wind is a current of air that moves horizontally through the surrounding air. Air currents can be vertical as well as horizontal.

curvature effect A consequence of the fact that the force binding water molecules to each other is strongest on a flat surface and weakens as curvature of the surface increases. Consequently, the saturation vapor pressure is higher over a curved water surface than over a flat one and increases as the curvature increases. Thus, water evaporates faster from a curved surface than from a flat one and small droplets evaporate much faster than big ones.

cutoff high A closed, middle latitude anticyclone that has moved into a higher latitude and become detached from the prevailing westerly air flow. This can cause blocking.

cutoff low A closed, middle latitude cyclone that has become detached from the prevailing westerly air flow and moved away into a lower latitude. This can cause blocking.

cutting-off The process by which a middle latitude anticyclone or cyclone is detached from the prevailing westerly air stream. This usually happens in the upper troposphere and produces slow-moving cutoff highs and cutoff lows around which weather systems are deflected.

cyclogenesis The series of events by which a cyclone develops in the polar front.

cyclolysis The weakening and disappearance of the cyclonic circulation of air that occurs as a cyclone family dissipates and high pressure comes to dominate.

cyclone (depression) An area of low atmospheric pressure around which there is a clearly defined wind pattern, the winds flowing cyclonically.

cyclone family A sequence of three or four mid-latitude frontal waves (depressions) that develop along the trailing edge of a very extended cold front. The first to form is called the primary and those following it are secondaries. Each secondary follows a track a little to the south of the one ahead of it.

cyclonic The adjective describing the direction in which air flows around a cyclone or trough. This is counterclockwise in the Northern Hemisphere and clockwise in the Southern Hemisphere.

cyclonic rain Precipitation associated with a cyclone (depression).

cyclonic shear Wind shear associated with vorticity around cyclones. Looking downwind, the winds are stronger on the right in the Northern Hemisphere and on the left in the Southern Hemisphere. This tends to set the air rotating cyclonically along the line of the wind.

cyclostrophic wind A strong, low-level wind that follows a very tightly curved path. When it blows around a very small, but intense, area of low surface pressure it can generate a dust devil. Winds around intense tropical cyclones are often cyclostrophic. The equation for calculating the speed (V) of the cyclostrophic wind is $V = \sqrt{\{(r/\rho)(\delta p/\delta x)\}}$, where r is the radius of the curved path, ρ is the air density, and $\delta p/\delta x$ is the pressure gradient.

dadur A northwesterly wind that blows down the valley of the Ganges River, India.

daily forecast A weather forecast for the period from 12 to 48 hours ahead.

daily mean The average value of a meteorological factor over 24 hours, counted from midnight to midnight. It is obtained by adding together the hourly readings and dividing the total by 24. In the case of temperature and pressure the average can be calculated as half the sum of the highest and lowest values recorded.

Dalton's law A law stating that in a mixture of gases, the total pressure is the sum of the pressures each component would exert if it alone occupied the same volume at the same temperature. This can be written as

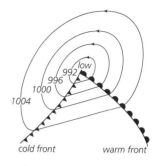

cyclone

$p = \Sigma_i \rho_i R_i T$, where p is the total pressure, Σ_i is the sum of the constituent gases, ρ_i is the density of each gas, R_i is the specific gas constant for each gas, and T is the absolute temperature.

damp air Air in which the relative humidity is higher than is usual for the place and season.

damp haze A fog that reduces visibility by an amount typical of a haze. The haze feels damp because it is caused by very small water droplets or hygroscopic nuclei.

damping A decrease in the amplitude of an oscillation that occurs over time because resistance to the oscillation drains energy from it.

damping depth The depth within which the temperature of soil is affected by diurnal or annual temperature changes at the surface. Assuming the soil to be homogeneous throughout the layer, the damping depth (D) is given by $D = (\kappa P/\pi)^{1/2}$, where κ is the thermal diffusivity of the soil and P is the period of the surface temperature change. The diurnal damping depth is 3 inches (7.9 cm) in dry sand and 5.5 inches (14 cm) in wet sand. The annual damping depth is 5 feet (1.5 m) in dry sand and 9 feet (2.7 m) in wet sand.

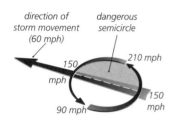

dangerous semicircle

dangerous semicircle The side of a tropical cyclone where the winds are strongest and tend to push ships into the path of the storm. The storm itself is moving, so its own speed of motion must be added to the wind speed on one side of the storm and deducted from it on the other side. The dangerous semicircle is on the side of the storm farthest from the equator.

Dansgaard–Oeschger event (DO event) A brief warm period during the most recent ice age, during which sea-surface temperatures in the North Pacific rose in a few decades by 5.4–9°F (3–5°C). When the event ended, the temperature fell just as abruptly to its ice-age level. There were many DO events, lasting from hundreds to several thousands of years.

Darcy's law A law describing the relationships among the factors that determine the rate at which groundwater moves through an aquifer. It states that $Q = kIA$, where Q is the rate of groundwater flow, k is the permeability and A the cross-sectional area of the aquifer, and I is the gradient of the slope down which the water is moving.

dark segment (Earth's shadow) A dark band sometimes seen just below the antitwilight arch shortly before sunrise or after sunset under conditions of haze.

dart leader A small lightning stroke that travels along the lightning channel, ionizing it, and is followed by the second major flash.

data buoy An instrument package that is located in a fixed position offshore and transmits continuous measurements of the surface conditions at sea.

datum (*pl.* data) (1) Something that is known or assumed to be true. (2) A premise from which inferences may be drawn. (3) The fixed starting point of a scale.

datum level A point or level surface that is used as a base from which other elevations can be measured. The most widely used datum level is the sea surface, from which altitude, the elevation of cities, height of mountains, and depth of the sea are measured.

day degrees (DD) A value calculated by multiplying together the number of days on which the mean temperature is above or below a particular datum level by the number of degrees by which it deviates from it.

dayglow Very weak light emitted in the mesosphere due to the bombardment of oxygen molecules by far ultraviolet radiation, at wavelengths below 200 nm. Dayglow becomes weaker as the Sun sets.

débâcle The breakup of river ice in spring.

declination The latitude of an astronomical body measured as the angle north (designated +) or south (–) of the celestial equator (a projection of the Earth's equator to the outermost limit of the universe).

deepening A fall in air pressure at the center of a cyclone.

Defense Meteorological Satellite Program (DMSP) A program run by the Air Force Space and Missile Systems Center to collect meteorological and oceanographic data and to monitor the space environment through which the Earth moves. This involves the design, building, launching, and maintenance of a number of satellites.

deflation The removal of material from the land surface by the action of the wind.

deglaciation The melting of an ice sheet or glacier, exposing the land beneath.

degree days The number of degrees by which the mean daily temperature is above the minimum temperature needed for growth (the zero temperature) for a particular crop plant.

degrees of frost An informal way to describe the number of degrees Fahrenheit by which the temperature is below freezing.

dehumidifier A device used to dry air that is too humid.

delta region Part of the atmosphere in which diffluence is occurring. The diverging streamlines make a triangular shape, reminiscent of the Greek letter Δ (delta).

dendrite A sharp spike, hexagonal in cross section, that projects from an ice crystal.

denoxification The removal of nitrogen oxides (NO_x) from stratospheric air by reactions that take place on the surface of particles in polar stratospheric clouds. Denoxification increases ozone depletion.

density The mass of a unit volume of a substance. It is measured in pounds per cubic foot (lb. ft.$^{-3}$) or pounds per cubic inch (lb. in.$^{-3}$), and in SI units in kilograms per cubic meter (kg m^{-3}). Under a standard atmosphere, pure water has a density of 0.62 lb. ft.$^{-3}$, 0.36 lb. in.$^{-3}$, or 10 kg m^{-3}. The density of air varies with altitude.

density altitude The height above the surface at which the density of the air has a specified value.

density current A flow of air caused by a difference between the density of the moving air and that of adjacent air.

density ratio The ratio of the density of the air at a specified altitude to the density of air at the same altitude in a standard atmosphere.

denudation The stripping away of the material that covers the surface of the land, exposing bare rock.

departure The extent to which the value of a meteorological factor differs from the mean value.

deposition The formation of ice on a solid surface by the direct conversion of water vapor into ice without passing through the liquid phase.

depression A midlatitude frontal cyclone. The term refers to a well-defined area of low atmospheric pressure; it is the air pressure that is "depressed."

depression family A series of wave depressions that move from west to east, carried by the prevailing westerlies. Each member of the family is linked to a wave in the jet stream above it. Depression families bring prolonged periods with gray skies and wet weather, interrupted only briefly by the ridges between one depression and the next.

depth-duration-area value (DDA value) The average depth of the precipitation that falls during a specified time over a specified area.

depth hoar (sugar snow) A layer of frost that forms by deposition just beneath the surface of a layer of snow.

descriptive climatology Climatology that is presented as descriptions of climates, using verbal accounts, graphs, tables, and other illustrations, but omitting discussions of the causes of phenomena or of climatological theory.

descriptive meteorology (aerography) Meteorology in which the composition and structure of the atmosphere and atmospheric phenomena are described, using verbal accounts, graphs, tables, and other illustrations, but the causes of those phenomena and meteorological theory are omitted.

desertification The deterioration of land until it has the characteristics of a dry desert.

desert wind A very dry wind that blows off a desert. It is hot in summer and cold in winter.

desiccation A long-term decrease in the amount of surface and/or ground-water in a region as a consequence of climatic change.

desorption The release of a gas that had previously been held in or on the surface of another substance.

determinism The idea that comprehensible natural laws govern the transition of a system from its present to a future state. That is, laws acting upon the present condition determine the future condition.

detrainment An outflow from a body of moving air into the surrounding air.

devastating drought A particularly severe contingent drought that occurs in summer and causes plants to wilt and die.

development The generation of vertical motion by buoyancy forces and the rise of warm air.

dew Moisture that condenses from the air onto the surface, most commonly onto plant leaves. It forms during cool nights when there is little wind.

dewbow A very faint rainbow sometimes seen in dew drops on the ground.

dew cell (dewpoint hygrometer) A hygrometer that measures the dewpoint temperature directly.

dew gauge (surface wetness gauge) An instrument used to measure dew. It comprises a styrofoam ball of a standard size held at the end of a vertical arm connected to a system of balances. The weight of the ball changes as dew condenses onto it and the instrument makes a

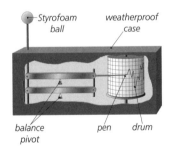

dew gauge

continuous record of this change by means of a pen at the end of one of the balance arms and a chart fixed to a rotating drum. The styrofoam ball is exposed, but the remainder of the device is enclosed by a weatherproof case.

dewpoint depression The difference between the ambient and dewpoint temperatures. This is approximately double the wet-bulb depression and is one of the ways of specifying the amount of water vapor present in the air.

dewpoint lapse rate The rate at which the dewpoint temperature decreases with height. It is about 0.55°F per thousand feet (1°C km^{-1}).

dewpoint temperature The temperature at which a parcel of air would become saturated if it were cooled with no change in the amount of moisture it contained or in the atmospheric pressure.

dew pond A shallow pond excavated by a farmer to provide water for livestock. Dew collects at the bottom of the pond, hence the name, but most of the water arrives as rain.

diabatic temperature change A change in the temperature of air due to contact with the surroundings.

diamond dust The popular name for ice prisms.

diathermancy (*adj.* diathermanous) The property of being transparent to radiant heat.

differential heating The warming of surfaces at varying rates when they are all equally exposed to sunshine, due to variations in the properties of the materials.

diffluence A flow of air in which two or more streamlines move away from each other.

diffraction The bending of light as it passes close to the sharply defined edge of an object by an amount proportional to the wavelength of the light (red light is diffracted more than blue light because its wavelength is greater). This causes the edge to appear blurred, but it also gives rise to interference, as different parts of the spectrum meet.

diffusion Mixing that occurs when one fluid is added to another, but the mixture is not stirred, shaken, or otherwise agitated. It is due to the random movement of molecules.

diffusion diagram A graphic representation of several of the processes involved in diffusion, such as the mean free path or mixing length of molecules together with their velocity.

diffusive equilibrium A condition in which the downward movement of air molecules due to gravity is balanced by the upward movement of molecules drifting into a region containing fewer molecules per unit volume.

diffusograph An instrument that measures diffuse radiation from the sky.

digital image A picture made from a continuously varying stream of data from an orbiting satellite or other remote source. The continuous variation is converted into discrete variation, in which the data change in small steps. Each discrete change is then given a numerical value as a picture element (conventionally abbreviated to "pixel") and assigned a location defined by coordinates.

dimethyl sulfide (DMS) A chemical compound, $(CH_3)_2S$, produced by the decomposition of dimethylsulfoniopropionate, a substance present in the cells of many single-celled marine algae. Some DMS escapes into the air, where it is rapidly oxidized. One product of DMS oxidation is sulfuric acid (H_2SO_4). Small droplets of H_2SO_4 form ideal cloud condensation nuclei and DMS is believed to be the principal source of such nuclei over the open ocean, where the air is almost completely free of dust.

direct cell One of the cells driven by convection and forming part of the general circulation of the atmosphere. The Hadley and polar cells are direct cells.

direct circulation The situation in which warm air is rising and cold air subsiding within a frontal zone.

dishpan experiments Laboratory experiments that simulate the general circulation of the atmosphere and demonstrate the importance of the horizontal transport of heat and momentum. The dishpan is shallow, shaped like a doughnut, and filled with water to which drops of dye are added. The center is kept cold while heat is applied around the outer edge, and the dishpan rotates. As currents develop in the water their directions are traced by the dye.

dispersion The separation of the component wavelengths as white light passes through a prism. Each wavelength is then seen as a band of a distinct color.

distorted water Water that has a different molecular structure from that of the main body of water of which it forms a part.

distrail (dissipation trail) A line of clear air in thin cloud behind an aircraft, caused by vortices in the wake of the aircraft.

disturbance Any small-scale variation in the general state of the atmosphere at a particular time and place.

disturbance line A weather system that occurs in spring and fall in West Africa. Moist air flowing from the southwest as part of the developing or fading monsoon circulation is overrun by dry air from the Sahara, producing a squall line several hundred miles long that travels westward at about 30 mph (50 kmh^{-1}). It dissipates once it has crossed the coast and encounters the cold water of the Atlantic. Disturbance lines are a major cause of rainfall between April and June and September and November.

diurnal range The difference between the daytime and nighttime temperature for a particular place. The mean range is between average temperatures and the absolute range is between the highest and lowest temperatures recorded.

divergence A flow of air in which streamlines move outward from an area of high pressure. This decreases the quantity of air in that area, and therefore decreases the atmospheric pressure.

Dobson spectrophotometer An instrument that measures the intensity of different wavelengths of ultraviolet radiation, from which the concentration of ozone present in the atmosphere can be inferred.

Dobson unit (DU) A unit of measurement used to report the concentration of a gas that is present in the atmosphere or in a particular part of the atmosphere. It refers to the thickness of the layer that gas would form if all the other atmospheric gases were removed and the gas in question were brought to sea level and subjected to standard sea-level pressure. The amount of ozone present in the stratospheric ozone layer is usually reported in Dobson units. One DU of ozone corresponds to a thickness of 0.01 mm (0.0004 inch) and the amount of ozone in the ozone layer is typically 220–460 DU, corresponding to a layer 2.2–4.6 mm (0.09–0.18 inch) thick.

dog days July and the first half of August, which is the hottest part of the summer in the Northern Hemisphere and when Sirius, the "dog star" and the brightest star, rises in conjunction with the Sun.

doldrums A sea area in which the winds are light and variable. The extent of the doldrums varies considerably with the seasons, but they are located in the intertropical convergence zone, on the side nearest the equator of the region in which the trade winds originate.

Doppler effect The rise in pitch of a sound that is approaching rapidly and the fall in pitch of a sound that is retreating rapidly. It was first tested

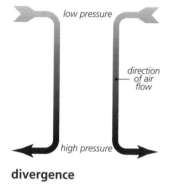

low pressure

direction of air flow

high pressure

divergence

using sound, but occurs with any form of wave radiation. As the source of emission approaches, the frequency increases. If the source is retreating, the frequency decreases. An increase in the frequency of a light beam is detected as a shift in the color toward the blue end of the spectrum. With a decrease in frequency, light is shifted toward the red end of the spectrum.

Doppler radar A technique in which two radar devices measure the Doppler effect on water droplets to determine the speed at which the droplets are rotating about a vertical axis. Doppler radar is employed to study air movements and measure wind speed inside tornadoes.

downburst A downdraft from a convection cell in a cumulonimbus cloud that reaches the surface and spreads to the sides, producing strong gusts and wind shear. In extreme cases a downburst from a supercell cloud can produce gusts of more than 75 mph (121 kmh^{-1}).

downdraft A current of sinking air inside a cumulonimbus cloud. Downdrafts usually travel at less than about 11 mph (18 kmh^{-1}), but can be much stronger in a supercell cloud.

downrush The strong downdraft that occurs during the cloudburst caused by the dissipation of a large cumulonimbus cloud. It comprises cold air dragged down by the falling snowflakes and raindrops.

downwash The transport to the surface of air caught in an eddy on the lee side of a hill or building. If the air is polluted, the pollutant will be brought to ground level.

drag The retarding effect that is caused by friction when air crosses a rough surface.

driven snow Snow that has been transported by the wind and deposited in snow drifts.

drizzle Liquid precipitation in which the droplets are smaller than 0.02 inch (0.5 mm) in diameter, are all of approximately similar size, and are very close together.

dropsonde An instrument package attached to a parachute that is dropped from an airplane. As it falls, its radio transmits data on pressure, temperature, and humidity recorded by its instruments at various altitudes.

dropwindsonde A dropsonde that can be tracked by means of the Global Positioning System to provide information on the wind speed and direction at different altitudes. Dropwindsondes are especially useful in studying tropical cyclones.

drought A prolonged period during which the amount of precipitation falling over a particular area is markedly less than the usual amount that falls over that period in that place. A drought is longer than a dry spell, but the length of time needed to define a drought varies greatly from place to place.

dry adiabat The continuous rate at which the temperature of dry air changes adiabatically with height.

dry adiabatic lapse rate (DALR) The rate at which rising unsaturated air cools adiabatically as it rises and warms as it descends. The DALR is 5.4°F per 1,000 feet (9.8°C km⁻¹).

dry airstream A flow of air that develops in the middle and upper troposphere behind a midlatitude cyclone and descends once the cyclone has passed.

dry bulb temperature The temperature registered by a thermometer, the bulb of which is dry and directly exposed to the air.

dry climate A climate in which the average annual precipitation is less than the potential evapotranspiration and plant growth is restricted by the lack of moisture.

dry deposition The transfer of particles from dry air to a surface onto which they are adsorbed.

dry haze Haze that contains no water droplets.

dry ice Solid carbon dioxide (CO_2), which, at standard sea-level pressure, must be kept at a temperature below −109.3°F (−78.5°C, 194.7K), the temperature at which it sublimes.

dry line (dewpoint front) A boundary that often forms over the Great Plains in spring and summer between hot, dry air to the west and warm, moist air to the east. Advancing dry air lifts the moist air ahead of it, often producing huge cumulonimbus clouds. Storms associated with dry lines frequently become tornadic.

dry season The time of year when precipitation is much lower than at other times. The dry season may occur in winter or in summer. Dry summers occur around the Mediterranean Sea and on the western sides of continents in latitudes 30–45° N and S. In regions with a monsoon climate, it is the winter that is dry.

dry snow Snow consisting of ice crystals with no liquid water between them. The individual crystals are joined to each other directly or by necks of ice.

dry spell A period of shorter duration than a drought during which no rain falls. In the United States a dry spell occurs if no measurable precipitation falls during a period of not less than two weeks.

dry tongue A tongue-shaped extension of dry air protruding into a region of moister air.

duplicatus A variety of clouds comprising layers, sheets, or patches of cloud that are at different heights but merge or overlap each other as seen from the ground.

düsenwind A strong northeasterly wind that blows through the Dardanelles, the strait in Turkey linking the Sea of Marmara with the Aegean Sea, when there is an area of high pressure over the Black Sea and lower pressure over the Aegean.

dust Solid particles that are lifted into the air by the wind, ejected as ash during volcanic eruptions, or released by industrial processes. Deserts are the main source of dust, but farming also contributes a large amount. The particles that become airborne most readily are about 40 μm in diameter. Sand dunes are accumulations of wind-blown sand, and fine soil is deposited as loess, which is very fertile.

Dust Bowl A region of the Great Plains in the United States, covering about 150,000 square miles (388,500 km^2) in southwestern Kansas, southeastern Colorado, northeastern and southeastern New Mexico, and the panhandles of Oklahoma and Texas, that experienced a severe drought from 1933 until the winter of 1940–41.

dust burden The weight of dust suspended in a volume of gas, measured in grams of dust per cubic meter at standard temperature and pressure.

dust collector A device that removes dust from industrial waste gases.

dust devil A twisting wind that resembles a tornado, but is much smaller, a great deal less violent, and not associated with a violent storm.

dust dome The increased concentration of dust particles that occurs beneath an urban dome.

dust horizon The upper surface of air held beneath an inversion that is visible from a distance because of the large amount of dust it holds. It resembles a horizon when seen partly silhouetted against the sky.

dust storm A strong wind that blows across bare ground in an arid region, lifting dust and keeping it aloft.

dust whirl (dancing devil, desert devil, sand auger, sand devil) A small column of rapidly rotating air that carries dust, sand grains, leaves, scraps of paper, and other light material.

dynamic climatology The scientific study of air movements and the thermodynamic processes that cause them.

dynamic meteorology The scientific study of atmospheric motion that predicts the future state of the atmosphere in terms of the physical variables of temperature, pressure, and velocity.

dynamic soaring A flying technique employed by some large sea birds. The bird glides downwind until it is very close to the sea surface. Then it turns into the wind to fly across the wind gradients behind wave crests or into the friction layer. In doing so it enters air that is moving more slowly than the air it leaves and this increases its speed in relation to the air (its airspeed), which increases the amount of lift generated by its wings, allowing it to climb back to its original height. In this way the bird can remain airborne for long periods with very little need to beat its wings.

Earth Observing System (EOS) The second part of Earth Science Enterprise, a program launched by NASA in 1991 with the aim of studying the solid Earth, oceans, and atmosphere as an integrated system. The first part of the program involved collecting and studying data from a number of satellites. EOS uses satellites dedicated to the task, the first of which was launched into polar orbit in 1999 and the second into a low-inclination orbit in 2000.

Earth Radiation Budget Experiment (ERBE) A study using radiometers carried on three satellites that measures all the solar radiation reaching the Earth and all the long-wave radiation leaving the surface. In the course of a month, the instruments provide measurements throughout almost the whole daily cycle for most regions of the Earth.

Earth Radiation Budget Satellite *(ERBS)* A satellite dedicated to the Earth Radiation Budget Experiment that was launched by the Space Shuttle *Challenger* in October 1984 and is operated by NASA. The *ERBS* also carries the Stratospheric Aerosol and Gas Experiment II.

Earth's shadow Another name for the dark segment.

Earthwatch Program A program to monitor changes in the environment that was established in 1973 and is managed by the United Nations Environment Program (UNEP), based in Geneva, Switzerland.

easterly jet A jet stream that blows in summer from east to west at a height of about 9 miles (15 km) and extends from the South China Sea to the southeastern Sahara.

easterly wave (African wave, tropical wave) A long, weak, low-pressure trough that moves from east to west across the tropical oceans, deflecting the trade winds and producing a wave pattern in the surface streamline. Easterly waves sometimes intensify to become tropical disturbances.

eccentricity The extent to which the orbital path of a planet or satellite deviates from a circle. If the geometric center of the ellipse its orbit describes is C, and the focus about which the body orbits is F_1, the distance between C and F_1 is the linear eccentricity, le. The location of the second focus, F_2, is at the center of the major axis, α, from C to the point at which the orbiting body is farthest from F_1. Eccentricity, e, is then given by: $e = le/\alpha$. At present the eccentricity of Earth is 0.017. It varies over a cycle of about 100,000 years from 0.001, which is almost circular, to 0.054.

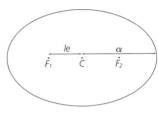

eccentricity

eddy Air that moves turbulently, its speed and direction changing rapidly and irregularly.

eddy correlation A technique to study the effect of the sea surface on the air immediately above it by comparing the mean speed and direction of air movement with the fluctuations in the vertical and horizontal components of that movement.

eddy diffusion The transport of heat energy and momentum by diffusion in air that is flowing turbulently. The rate at which diffusion occurs is known as the eddy diffusivity and is measured in units of area per second.

eddy viscosity The diffusion of momentum in air that is moving as a turbulent eddy due to the mingling of air at the boundaries between eddies.

effective gust velocity The upward vertical component of a gust of wind that would produce a given acceleration on an aircraft flying straight and level through air of constant density at the recommended cruising speed of that aircraft.

effective precipitable water The proportion of the precipitable water that can fall as precipitation.

effective precipitation A value for the aridity of a climate that is calculated as r/t, where r is the mean annual precipitation in millimeters and t is the mean annual temperature in °C. Precipitation falling as snow must be converted to its rainfall equivalent. If r/t is less than 40 the climate is considered to be arid, and if r/t is greater than 160 the climate is perhumid (extremely wet).

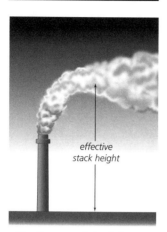

*effective
stack height*

effective stack height

effective stack height The height at which a chimney plume begins to move downwind after it has emerged from the top of a smokestack.

effective temperature The temperature of saturated air with an average wind speed of no more than 0.2 ms[-1] (0.45 mph), which would produce the same sensation of comfort in a sedentary person wearing ordinary indoor clothes as air with the movement, humidity, and temperature to which that person is actually exposed.

Ekman layer Part of the planetary boundary layer in which the wind blows at an angle across the isobars, the movement of air being balanced between the pressure gradient force, Coriolis effect, and friction.

Ekman spiral The spiral pattern that emerges if the changing direction of the wind with increasing height is plotted on a two-dimensional surface.

elastic modulus *(E)* The ratio of the force applied to a body as stress to the extent to which it deforms (strain). For air, $E = \gamma p$, where γ is calculated from the heat capacities of the principal constituents and p is the pressure.

El Chichón A volcano in Mexico that erupted from March to May 1982, ejecting gas that included up to 3.6 million tons (3.3 million tonnes) of sulfur dioxide, all of which was converted into sulfuric acid. By May 1 this cloud had circled the Earth. A year later the cloud had spread to cover almost the whole of the Northern Hemisphere and a large part of the Southern Hemisphere. It produced a marked warming of the lower stratosphere and in June it reduced the average global temperature by about 0.4°F (0.2°C).

electrically scanning microwave radiometer (ESMR) A radiometer that was launched on the Nimbus 5 satellite in December 1972 and continued to function until the end of 1976. Its purpose was to monitor sea ice.

electric field A region in which a force is exerted on any electrically charged body that enters it.

electrification ice nucleus A fragment of ice produced when a dendrite shatters in a strong electric field. Such fields occur in cumulonimbus clouds.

electrogram A record showing changes over time in the atmospheric electric field.

electrometer An instrument used to measure the atmospheric electric field.

electron capture detector An instrument that measures extremely small concentrations of substances in the air. It is most sensitive to bromine, chlorine, fluorine, and iodine, making it very suitable for detecting organochlorine pesticides, such as DDT, and CFCs.

electrostatic charge A positive or negative electric charge that is at rest (it does not flow).

Elektro satellite (Geostationary Operational Meteorological Satellite) A Russian meteorological system launched into a geostationary orbit above 76.83° E on October 31, 1994. It monitors atmospheric processes, sea-surface temperature, and wind velocity at various altitudes; detects natural hazards; and measures variations in the Earth's magnetic field and solar ultraviolet and X-ray radiation. It also transmits pictures taken in visible and infrared light.

El Niño A change in the prevailing winds over the equatorial South Pacific that occurs at intervals of two to seven years. Ordinarily, these are from the southeast, driving a surface current that carries warm water westward. During an El Niño, the winds weaken or may even cease to blow or reverse direction. The wind-driven surface ocean current ceases or reverses and warm water accumulates off the coast of South America. El Niño brings rain to the parched coastal strip at about Christmas, which is how the phenomenon earned its name of "(male) child" (i.e., the Christ child).

Elsasser's radiation chart A chart used to determine the components that together make up the outgoing radiation from the Earth's surface.

elvegust (sno) A cold, descending squall in the upper parts of Norwegian fjords.

embata An onshore wind from the southwest that blows across the Canary Islands.

emissary sky A sky covered by patchy cirrus cloud. This acts as an emissary of the rain and winds that are to come.

emissivity The amount of radiation a body emits, expressed as a proportion of the radiation that would be emitted at the same wavelength by a black body at the same temperature. The emissivity of gases varies considerably with wavelength, but that of solid objects remains fairly constant.

energy balance model (EBM) A model of the atmosphere that describes the change in sea-level temperature with latitude and the balance of energy input and output in each latitudinal belt. The relationship it describes is summarized by $\rho C \Delta T(\theta) \div \Delta T = R\!\downarrow\!(\theta) - R\!\uparrow\!(\theta) +$ transport into belt θ; ρC is the heat capacity of the area being studied,

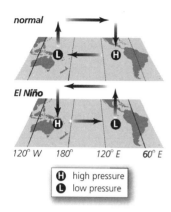

normal

El Niño

120° W   180°   120° E   60° E

⊕ high pressure
⊙ low pressure

ENSO

ΔT is the change in temperature; θ is the latitudinal belt; $R\downarrow$ is the radiant energy entering the system; and $R\uparrow$ is the infrared radiation leaving the system.

energy exchange The warming of a body that comes into contact with another body at a higher temperature.

ENSO (El Niño–Southern Oscillation event) A change in the distribution of surface pressure and wind direction that affects the equatorial South Pacific at intervals of two to seven years, producing El Niño events.

enthalpy The heat that can be felt (sensible heat) and that is transferred between bodies at different temperatures.

entrainment Mixing that takes place between a body of air and the air surrounding it.

entrance region The area in which air is being drawn toward the core of a jet stream.

entropy A measure of the amount of disorder that is present in a system.

envelope orography A technique sometimes employed in mathematical models used in weather forecasting. It assumes the valleys and passes in mountain ranges are filled mainly with stagnant air. This allows them to be ignored, effectively increasing the average height of the mountains.

environmental lapse rate (ELR) The rate at which the air temperature decreases with height as this is measured at a particular time and place. The ELR is not the result of adiabatic cooling, but simply the actual temperature that is observed, and it is very variable.

eolian (aeolian) Transported by the wind.

equation of motion The mathematical equation describing the movement of air according to Newton's second law of motion. It is most often written as $a = F/M$, where a is acceleration and F is the force acting on a body of air with a mass M. The equation assumes that the mass of a parcel of air remains constant. Therefore, it cannot be applied to the development of a cumuliform cloud in which the mass changes as a consequence of entrainment.

equation of state (ideal gas law) An equation or law relating the temperature, pressure, and volume of an ideal gas. The basic equation is $pV = R^*T$, where p is the pressure, V the volume, T the temperature, and R^* is the universal gas constant (8.31434 JK^{-1} mol^{-1}). Air is unconfined and thus has no precise volume, so the equation used in meteorology substitutes density for volume. It is then $p = \rho RT$, where ρ is the density and R is the specific gas constant for air (287 $JK^{-1} kg^{-1}$).

equatorial air Warm, humid air that forms an air mass covering the equatorial belt in both hemispheres.

equatorial climate The climate of the region bounded approximately by latitudes 10° N and 10° S. It is warm and humid throughout the year, with little seasonal variation.

equatorial easterlies The northeasterly and southeasterly trade winds during the summer, when the westerly winds blowing above them are either nonexistent or too weak to affect the lower troposphere.

equatorial trough A wide belt of low surface pressure that encircles the Earth where the trade winds of both hemispheres converge in the intertropical convergence zone, causing air to rise.

equatorial vortex The vortex at the center of the cyclonic circulation produced by an equatorial wave over the Pacific Ocean. The vortex travels westward toward the Philippines, but rarely intensifies into a tropical cyclone.

equatorial wave A wave disturbance in the equatorial trough over the Pacific Ocean when the trough is far enough from the equator for the Coriolis effect to generate cyclonic motion.

equatorial westerlies Westerly winds that occur in summer between the northeasterly and southeasterly trade winds. They are most strongly evident over continents, especially over Africa and southern Asia.

equilibrium level (level of zero buoyancy) The height at which rising air becomes neutrally buoyant and will rise no farther.

equilibrium vapor The water vapor in the very shallow layer of air in contact with an exposed surface of liquid water. The layer of air is about 0.04 inch (1 mm) thick, and the water vapor it contains is in equilibrium because its amount does not change.

equinoctial gale A wind storm that occurs around the time of an equinox. There is no basis for a folk belief that gales are more common at the equinoxes.

equinoctial rain Rainy seasons that occur in some parts of the Tropics at around the equinoxes.

equinox One of the two dates each year when at noon the Sun is directly overhead at the equator and day and night are of equal length. The equinoxes fall on March 20–21 and September 22–23.

equivalent-barotropic model A model of the atmosphere in which it is assumed that wind direction does not change with height, all the contours on any isobaric surface are parallel, and vertical movements are presumed to be equivalent to those at an intermediate level, known as the equivalent-barotropic level.

equivalent potential temperature The temperature a parcel of air would have if it were decompressed at the saturated adiabatic lapse rate (SALR) almost to zero pressure and then recompressed at the dry adiabatic lapse rate (DALR) to 1,000 mb (sea-level pressure).

equivalent temperature (isobaric equivalent temperature) The temperature a parcel of air would have if all its water vapor condensed, the latent heat of condensation warmed the air, and the pressure remained constant. "Equivalent temperature" is sometimes used as a synonym of equivalent potential temperature.

etalon An interferometer that comprises half-silvered, plane-parallel glass or quartz plates a fixed distance apart, with a film of air enclosed between them.

etesian Cool, dry northerly winds that blow over the eastern Mediterranean. They are most frequent from May to October and can be so strong it is impossible for sailing vessels to travel against them.

Eumetsat The European Organization for the Exploitation of Meteorological Satellites, an intergovernmental organization founded in 1986 that establishes and maintains operational meteorological satellites on behalf of Austria, Belgium, Denmark, Germany, Finland, France, Greece, Ireland, Italy, the Netherlands, Norway, Portugal, Spain, Sweden, Switzerland, Turkey, and the United Kingdom.

Eurasian high An area of high pressure that develops across Eurasia in winter and collapses in April.

eustasy The worldwide change in sea level due either to tectonic movements of the Earth's crust or to the expansion or melting of glaciers.

evaporation The change from the liquid to the gaseous phase.

evaporation pan A device for measuring the rate of evaporation, comprising a container of a standard size holding water that is exposed to the air.

evaporation pond An enclosed body of sea water that is allowed to evaporate in order to obtain the salts that crystallize from it.

evaporative power (evaporative capacity, evaporativity, potential evaporation) The ability of the climate of a region, or the weather at a particular time, to evaporate water. It is measured as the rate of evaporation from the surface of chemically pure water at the temperature of the air in contact with it.

evaporimeter A simple instrument for measuring the rate of evaporation. There are several designs. One consists of a graduated reservoir, sealed with a cork and with a ring from which it can be hung. At the bottom of the reservoir a U-tube expands to an open surface of known area covered with filter paper. As water evaporates from the filter paper, the level falls in the reservoir. Scientists use a lysimeter for more precise measurements.

evapotranspiration The loss of water from the surface due to the combined effects of evaporation and transpiration from plants.

evapotranspirometer An instrument used to measure potential evapotranspiration, which is an important value in calculations connected with the Thornthwaite climate classification.

evapotron An instrument that measures the extent and direction of eddies involved in the vertical transfer of water vapor. This makes it possible to measure directly the rate of evaporation over a very short period.

exitance A measure of the amount of electromagnetic radiation released from a unit area of a surface.

exit region The area in which air is being expelled from the core of a jet stream.

exosphere The layer of the atmosphere between about 300 and 450 miles (480–725 km) above the surface. The predominant gases are atomic oxygen, helium, and hydrogen, about 1% of which are ionized.

expansional cooling The mechanism by which an expanding fluid cools. As they move farther apart, the molecules in one body of fluid (the expanding body) transfer kinetic energy to outside molecules (in the surrounding fluid). Consequently, energy leaves the expanding body and it cools.

exponential An adjective describing a function that varies as the power of a particular quantity. If $x = y^a$, x is said to vary exponentially with a. An exponential change is one in which a quantity changes by a constant proportion that is calculated in each period on the accumulated total of the changes in previous periods.

exposure The extent to which a site experiences the full effect of such meteorological events as wind, sunshine, frost, and precipitation.

extratropical cyclone A region of low pressure (depression) in middle latitudes (outside the Tropics) around which air circulates cyclonically.

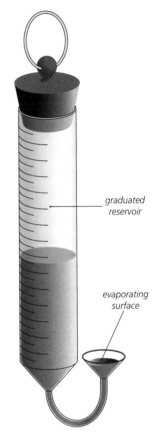

graduated
reservoir

evaporating
surface

evaporimeter

extratropical hurricane A severe storm that occurs in the Arctic or Antarctic.

extreme The highest or lowest temperature observed at a particular place over a specified period.

eye of storm The region of calm air that lies at the center of a deep cyclone.

eye of wind The direction from which the wind is blowing or the point on the horizon from which it appears to blow.

eyewall The circle of towering cumulonimbus clouds that surrounds the calm center of a tropical cyclone or extratropical hurricane.

facsimile chart A weather chart that is distributed as a fax from a central meteorological office.

facsimile recorder The device that transmits a facsimile chart.

Fahrenheit temperature scale The temperature scale on which pure water freezes at 32° and boils at 212°, and average body temperature is 98.6°. This scale is used in Britain and the United States, although it is being replaced by the Celsius temperature scale, which is always used in scientific publications.

fair An adjective describing weather that is pleasant for a particular place at a particular time of year. This generally implies light winds, no precipitation, and less than half the sky covered by cloud.

fair weather cumulus Small, white, fleecy clouds that appear in fine weather and that deliver no precipitation.

fair weather electric field The electric field that exists in the atmosphere and at the ground and sea surface.

fallout The removal from the air of solid particles that descend to the surface by gravity.

fallout front The lower boundary of an area of fallout, where solid particles descending by gravity reach the surface.

fallout wind A wind carrying particles that are at the same time descending by gravity.

fall speed The speed with which a body falls through the air, equal to the terminal velocity of the body minus the velocity of any upward air current to which it is exposed.

fallstreak hole A hole that sometimes develops in clouds composed of supercooled water droplets. Droplets in part of the cloud freeze and then grow into raindrops, which fall from the cloud, often in the form of virga (fallstreaks), leaving behind clear air from which the droplets have been removed.

false cirrus Cloud resembling cirrus that forms from the upper part of a cumulonimbus cloud that has dissipated or from which it has become detached.

false color Colors used in satellite images that differ from the actual colors of the surfaces they represent. In many false-color images infrared radiation appears as visible red. Vegetation is highly reflective to infrared wavelengths and so it often appears red in false color pictures.

fanning The pattern made by a chimney plume when the gases and particles travel downwind smoothly and horizontally without dispersing.

far-infrared radiation Infrared radiation with a wavelength greater than 15 μm.

fata morgana A mirage seen when a thin layer of air lies above a cold surface, most commonly of the sea, and refraction produces a greatly magnified superior image of distant buildings or cliffs that sometimes resemble castles partly in the sky and partly beneath the sea. The phenomenon is especially common in the Strait of Messina, between the Italian mainland and Sicily, a site associated with the mythical submarine palace of Morgan le Fay, or Fata Morgana, the legendary sister of King Arthur.

fern frost Frost that forms in patterns resembling fern fronds. It is most often seen early in the morning on windows and is a variety of frozen dew.

Ferrel cell In the three-cell model of atmospheric circulation, the indirect cell between the Hadley and the polar cells in each hemisphere. Air moves through the Ferrel cell in the opposite direction to its movement through the two direct cells, rising at the polar front and subsiding in the subtropics.

fetch The distance air moves across the surface of the sea or ocean.

fetch effect An alternative name for the leading-edge effect that occurs when air passes from one type of surface to another. The depth of the internal boundary layer of air affected by the new surface increases with distance (fetch) from the leading edge.

fibratus (fib) A species of clouds consisting of long filaments that are almost straight or irregularly curved and do not end in hooks.

fibril A trail of cloud sometimes seen extending from a cumulonimbus cloud.

fanning

fiducial point A fixed position from which other positions can be measured. The fiducial point (or standard temperature) of a barometer is the temperature at which that barometer gives a correct reading at latitude 45°. At any other temperature or latitude the barometer reading must be adjusted. The fixed point marking zero on the scale of a Fortin barometer is also called the fiducial point.

field capacity The amount of water a particular soil will retain under conditions that allow water to drain freely from it.

field changes The rapid fluctuations in the vertical component of the electrical field that occur near the surface during a thunderstorm.

filling An increase in air pressure at the center of a cyclone.

fire storm A wind storm caused by an intensely hot fire.

fire weather Weather conditions that favor the ignition and spread of forest fires in a specified area.

firn Snow that lies on the ground for a second winter, having failed to melt during the intervening summer.

firn line (annual snow-line, firn limit) A line on a glacier marking the highest elevation at which snow falling in winter melts during the summer.

firn wind (glacier wind) A katabatic wind that blows from a glacier during the day.

first gust A sudden sharp increase in the wind speed as a cumulonimbus cloud enters its mature stage. It is caused by the arrival of the cold downdraft at the surface.

first-order station Any U.S. weather station staffed wholly or partly by National Weather Service personnel.

fitness figure (fitness number) A value, used in Britain, for the suitability (fitness) of weather conditions for the safe landing of aircraft at an airport.

flaschenblitz Lightning that flashes upward from the top of a cumulonimbus cloud.

flash frost A frost that appears on roads very suddenly soon after dawn.

flaw An old English name for a sudden squall of wind.

floccus (flo) A species of cloud consisting of cumuliform elements with a base that is more or less ragged. Floccus often occurs with virga (fallstreaks).

Flohn classification A climate classification based on the influence of air masses and prevailing winds. It is considered one of the best schemes of this type and especially useful as an introductory outline of climate classification.

flurry A sudden, brief shower of snow accompanied by a gust of wind, or a mild wind squall, even if it brings no snow.

flux The rate at which a fluid or radiation flows across an area.

fog Precipitation in the form of stratus cloud that extends to the ground or sea surface and reduces horizontal visibility to less than 1,094 yards (1 km).

fog bank A well-defined mass of fog seen from a distance, especially at sea.

fogbow A white rainbow seen in fog.

fog dispersal The deliberate clearance of fog in order to increase horizontal visibility sufficiently to permit aircraft and vehicles to maneuver on the ground and vessels to move safely in coastal waters and harbors.

fog drip Water deposited on trees and other tall structures by fog that drips to the ground. This can deliver as much water to the ground as a light shower.

fog droplet An individual fog particle, consisting of a water droplet 0.00004–0.0008 inch (1–20 μm) in diameter. A typical fog contains less than 0.001 ounce of water per cubic foot (1 gm^{-3}).

fog horizon The boundary between the sky and the upper surface of a fog layer trapped beneath an inversion. Seen from above the fog, the boundary resembles the true horizon.

Fog Investigation Dispersal Operations A method devised during World War II to clear fog from military airfields. Gasoline vapor was pumped through perforated metal pipes laid on either side of the runway and ignited at the perforations. It burned with a fierce heat that raised the dewpoint temperature, causing the fog to evaporate.

fog shower A type of precipitation that can occur above the lifting condensation level on a mountain engulfed by a passing cumulonimbus or large cumulus cloud containing supercooled water droplets that freeze into rime frost or glaze on contact with small objects. The cloud appears as fog to an observer on the mountainside and the impact of the very cold droplets feels like a shower of rain.

fog streamer A wisp of fog near the surface that forms when cold air blows across a lake.

föhn air Warm air carried down a mountainside by a föhn wind.

föhn cloud A cloud, usually of the lenticularis type that forms on the lee side of a mountain range and is often associated with a föhn wind.

föhn cyclone A cyclone on the lee side of a mountain that draws air down the mountainside, creating a föhn wind.

föhn island A low-lying area of warm air on the lee side of a mountain affected by a föhn wind.

föhn nose The characteristic shape of the isobars that indicates a fully developed föhn wind on a synoptic chart.

föhn pause The boundary between föhn air and the cold air adjacent to it.

föhn period The length of time during which a specified place lies beneath föhn air.

föhn phase The stage that has been reached in the development of a föhn wind when this is due to blocking by an inversion at the level of the mountain summit on the windward side.

föhn wall The upper surface of the cap cloud blanketing a mountain peak, seen from the lee side of the mountain.

föhn winds Warm, dry winds that occur on the northern side of the European Alps, most commonly in spring. They are of the same type as the North American chinook wind and also occur on the eastern side of the New Zealand Alps and the leeward side of the mountains of the Caucasus and in Central Asia.

following wind A tailwind, or a wind blowing in the same direction as waves moving over the sea surface.

forced convection Convection that occurs when air at the same temperature as the surrounding air, and therefore having neutral buoyancy, is made to rise or sink.

forecast period The length of time covered by a weather forecast.

forecast-reversal test A test to measure the usefulness of a method for forecast verification. The same verification method is applied simultaneously to two weather forecasts, one an actual forecast and the other a fabricated forecast predicting the opposite conditions. Each forecast is given an accuracy score and the scores are compared. This evaluates the verification test, because the real forecast should achieve a markedly higher score for accuracy than the fabricated one.

forecast skill A measure of the accuracy of a weather forecast on a scale ranging from 0 (completely wrong) to 1 (completely correct).

forecast verification Any technique used to measure the accuracy of a weather forecast.

forensic meteorology The branch of meteorology concerned with the relevance of atmospheric conditions to legal problems.

forked lightning Lightning flashes that are seen as brightly luminous, jagged lines between a cloud and the ground or between two clouds.

Fortin barometer A portable mercury barometer that is simple to use and sufficiently accurate for most purposes.

fossil turbulence Local variations in temperature and humidity, produced by turbulent flow, that persist after the movement causing them has ceased and the density of the air has become uniform.

fractocumulus The cloud species cumulus fractus (Cu_{fra}), comprising fragments of broken or ragged cloud that have been torn from cumulus or that are the remains of cumulus that has dissipated.

fractostratus The cloud species stratus fractus (St_{fra}) that remains as stratus dissipates.

fractus (fra) A species of clouds comprising fragments of cumulus or stratus that have been torn from the parent cloud or that remain after the parent cloud has dispersed.

fragmentation nucleus A tiny splinter of ice that snaps off from a larger ice crystal during a collision between crystals and then serves as an ice nucleus onto which a new ice crystal will grow.

Framework Convention on Climate Change An agreement reached at the United Nations Conference on Environment and Development held in Rio de Janeiro, Brazil, in June 1992 (the Rio Summit, or Earth Summit). The Framework Convention urges national governments to promote research into climate change and to reduce emissions of greenhouse gases. Its most direct achievement is the Kyoto Protocol that sets targets for reduced emissions.

frazil ice Ice crystals that form on the surface of the sea as the temperature falls below freezing. They make the water look oily, and as more of them form the sea becomes covered with slush.

free atmosphere The approximately 95% of the total mass of the atmosphere that lies above the planetary boundary layer.

free convection Convection caused by the warming of the air from below.

free radical A group of atoms with unpaired electrons. Free radicals are highly reactive.

freezing The change of phase from liquid to solid; in the case of water, the change from liquid to ice.

freezing drizzle Drizzle comprising supercooled droplets that freeze on contact with the ground.

freezing fog Fog that forms when the air temperature is below freezing. The low temperature chills surfaces to below freezing and fog droplets freeze onto them, coating surfaces with ice.

freezing index The cumulative number of degree days when the air temperature is below freezing. The indices are available for the entire world, broken into areas measuring 0.5° longitude by 0.5° latitude.

freezing level The lowest height above sea level at which the air temperature is 32°F (0°C).

freezing-level chart A synoptic chart that uses contour lines to show the height of the constant-temperature surface of the freezing level.

freezing nuclei Small particles onto which supercooled water droplets will freeze.

freezing rain Rain consisting of supercooled raindrops that freeze immediately on contact with the ground.

frequency The rate at which a regularly repeating event recurs. The frequency of a wave (f) is calculated by $f = c/\lambda$, where c is the speed of the wave and λ is the wavelength.

frequency-modulated radar A more advanced version of continuous-wave radar that can measure distance, because each part of the signal is tagged to make it and its reflection recognizable.

fresh breeze A force 5 wind on the Beaufort wind scale. It blows at 19–24 mph (31–37 kmh^{-1}).

fresh gale A force 8 wind on the Beaufort wind scale. It blows at 39–46 mph (63–74 kmh^{-1}).

fresh water Water that contains very little salt (usually less than 0.03% by volume). When water evaporates or freezes, only water molecules enter the air or form ice crystals. Consequently, precipitation consists of fresh water, although airborne salt crystals may dissolve in atmospheric water droplets. Ice, including sea ice and icebergs, is also made from fresh water, although small amounts of salt water may be held between ice crystals.

friagem A spell of cold, cloudy weather, with occasional rain, that occurs in winter in the middle and upper Amazon Basin and lasts four or five days.

friction The force that resists the motion of a solid body or fluid that is in contact with a surface that is stationary or moving at a different speed or in a different direction.

friction layer (surface boundary layer) The lowest approximately 10% of the depth of the planetary boundary layer, in which turbulent flow ensures thorough mixing of the air.

friction velocity (shear velocity, u^*) The velocity of air in the planetary boundary layer, above the laminar boundary layer; $u^* = (\tau/\rho)^{1/2}$, where τ is the tangential stress on the horizontal surface and ρ is the air density.

fringe region (spray region) The uppermost part of the exosphere, where the atmosphere is so rare that individual atoms seldom collide.

frog storm (whippoorwill storm) In North America, the first bad weather to follow a spell of warm weather in spring.

front The boundary between two air masses.

frontal analysis The study of weather charts in order to identify the boundaries between adjacent air masses and mark the fronts separating them.

frontal contour A line that marks the intersection between a front and a surface.

frontal cyclone A cyclone associated with a frontal system. The term is synonymous with frontal wave, but is sometimes used to distinguish a cyclone of this type from a tropical cyclone.

frontal fog (precipitation fog) Fog associated with a front, where warm air is being lifted above colder air.

frontal inversion A temperature inversion at a front, where warm air lies above cold air. The front forms a barrier to air that is rising convectively in the cold air mass beneath it.

frontal lifting The forced ascent of warm air as it rises over cold air at a warm front or as it is undercut by advancing cold air at a cold front.

frontal passage The movement of a front over a point on the surface.

frontal precipitation Precipitation that falls from clouds associated with a weather front.

frontal profile A diagram showing a vertical cross section through a front, sometimes with the clouds associated with the front.

frontal slope The gradient of a front, measured either as the angle between the front and the surface or as the ratio of vertical-to-horizontal

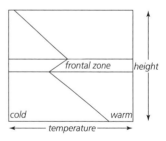

frontal inversion

distance. A warm front has an average slope of between $1/2$ ° and 1°, or between about 1:115 and 1:57. A cold front slopes much more steeply, at about 2°, or about 1:30.

frontal strip The representation of a front as two parallel lines on a weather map, rather than as a single line. This illustrates the fact that the front is a band of transition between two air masses, rather than the abrupt change suggested by a single line.

frontal structure The way air moves, clouds form, and precipitation develops in a frontal system.

frontal system A complete system of warm, cold, and occluded fronts as these are shown on a weather map.

frontal thunderstorm A thunderstorm that develops in warm, moist air made unstable by frontal lifting.

frontal wave (frontal depression) The wave that develops in the course of frontogenesis. Once the wave has formed, cyclonic circulation surrounds the center of an area of low pressure at the crest of the wave. This constitutes the depression. The system travels with a warm front at the leading edge of the wedge of warm air and a cold front at the trailing edge. The cold air travels faster than the warm air, causing the cold front to override the warm front. The tip of the wedge of warm air is lifted clear of the surface and the fronts appear on a weather map as an occlusion. When the whole of the wedge of warm air has been lifted above the surface, the system dissolves as the warm and cold air mix.

frontal zone The region of a front where the temperature gradient is strongest.

frontogenesis The formation and subsequent development of a boundary (front) between two air masses.

frontolysis (frontal decay) The dissolution and disappearance of a weather front that occurs when there is no longer any difference in the characteristics of two adjacent air masses.

frost A coating of ice crystals that forms on solid surfaces.

frost day A day on which frost occurs.

frost hollow (frost pocket) A small, sheltered, low-lying area in a hilly region that experiences freezing temperatures more frequently than the surrounding area, because at night cold air sinks down the hillsides and accumulates.

frostless zone The part of a hillside that remains free from frost on nights when frost forms in the valley. Cold, dense air that flows downhill by gravity is replaced by warmer air that keeps the upper hillside relatively warm.

frost point The temperature at which water vapor turns directly into ice.

frost smoke Steam fog consisting of ice crystals that forms when the temperature is well below freezing.

frozen fog Fog comprising low clouds consisting of ice crystals.

frozen precipitation Precipitation of any kind that reaches the ground in the form of ice.

Fujita tornado intensity scale A six-point scale (F-0 to F-5) used to classify tornadoes according to the damage they cause. Tornadoes are grouped as weak (F-0 and F-1), strong (F-2 and F-3), and violent (F-4 and F-5).

Fujiwara effect A phenomenon in which two typhoons of approximately similar size approach to within about 900 miles (1,450 km) of each other and begin to turn about a point approximately halfway between them. If one storm is much bigger than the other, they turn about a point closer to the larger storm, which then absorbs the smaller one.

fume A mass of solid particles, less than 0.00004 inch (1 μm) in diameter, suspended in the air, that result from the condensation of vapors, deposition, or chemical reactions.

fumigating The downwind pattern made by a chimney plume that widens and sinks with increasing distance from the smokestack. This brings the gases and particles to ground level, where they pollute the air.

fumulus A cloud layer so thin and tenuous as to be barely visible.

funnel cloud A funnel-shaped cloud that develops inside a mesocyclone and descends through the base of a cumulonimbus cloud. If it touches the ground it becomes a tornado.

fumigating

funneling An acceleration of the wind that occurs when it is forced through a narrow passage.

furiani A strong southwesterly wind that blows in the region near the Po River, Italy.

furious fifties The fierce gales encountered in the Southern Ocean between latitudes 50° S and 60° S.

gale A strong wind, ranging from force 7 to force 10 on the Beaufort wind scale. The wind speeds of the four gales are moderate (force 7), 32–38 mph (51.4–61.1 kmh^{-1}); fresh (force 8), 39–46 mph (62.7–74 kmh^{-1}); strong (force 9), 47–54 mph (75.6–86.8 kmh^{-1}); and whole (force 10), 55–63 mph (88.4–101.3 kmh^{-1}). A wind stronger than a whole gale is called a storm, and one weaker than a moderate gale is a strong breeze.

gallego A cold, northerly wind that blows across Spain and Portugal.

garúa (camanchaca, Peruvian dew) A wet mist or very fine drizzle that falls on the lower slopes of the Andes in Peru in winter, sometimes lasting for weeks.

gas constant A value used when the gas laws are combined into the equation of state. The universal gas constant (R^*) has a value of 8.314 J K^{-1} mol^{-1}. The specific gas constant (R) varies from one gas to another. $R = 10^3 R^*/M$, where M is the molecular weight of the gas (the value must be multiplied by 1,000 because moles are defined in grams and the unit of mass is the kilogram). Air is a mixture of gases. Adding together the constants for each of them gives $R = 287$ J kg^{-1} K^{-1} (17.3 cal pound^{-1} °F^{-1}) for dry air with a relative molecular mass of 29.0. This is only approximately true for moist air, because the presence of water vapor reduces the density of moist air to about 0.5% less than that of dry air at the same temperature and pressure. For precise calculations of conditions inside clouds, the use of the virtual temperature gives moist air the same gas constant as dry air.

gas laws The physical laws by which the temperature, pressure, and volume of an ideal gas are related. The laws can be combined into a single equation of state, pV = a constant, where p is the pressure and V the volume, from which a universal gas equation can be derived: $pV = nR^*T$, where n is the amount of gas in moles, T is the temperature, and R^* is the universal gas constant. If the specific gas constant for air (R) is substituted for the universal gas constant R^*, the law can be expressed as: $p = \rho RT$, where ρ is the density.

Gaussian distribution (normal distribution) In statistics, the way in which values of a variable quantity appear on a graph when beneath the graph curve there is an equal area to either side of the mean. The curve is bell-shaped and its maximum height marks the mean.

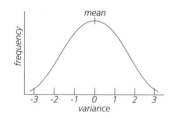

Gaussian distribution

gegenschein A faint glow sometimes seen as a circle or ellipse of light in the sky opposite the Sun.

gending A dry, southerly föhn wind that blows across the northern plains of Java.

general circulation All the movements of the atmosphere by which heat is transported away from the equator and into higher latitudes, winds are generated, and clouds and precipitation are produced.

general productivity model A climate model that interprets climatic conditions to predict agricultural production throughout the major farming regions of the world.

Geneva Convention on Long-Range Transboundary Air Pollution A legally binding international agreement that came into force in 1983. It lays down the general principles for international cooperation to reduce the emission of air pollutants that drift across international frontiers and also establishes an institutional framework for research and the development and implementation of policy.

genitus The emergence of a new cloud from only one part of a mother cloud.

Genoa-type depression A type of depression that is common in winter over the western Mediterranean, following a sudden drop in pressure around October 20, when the Azores high collapses.

gentle breeze A wind of force 3 on the Beaufort wind scale, blowing at 8–12 mph (13–19 kmh^{-1}).

geopotential height A value for height above mean sea level that takes account of the increase in gravitational acceleration with latitude.

Geostationary Earth Radiation Budget (GERB) An experiment carried on Meteosat MSG satellites to observe and measure the radiation reflected and emitted by the Earth.

Geostationary Operational Environmental Satellite *(GOES)* A series of these U.S. weather satellites transmit data to the National Oceanic and Atmospheric Administration (NOAA) receiving station at Wallops, Virginia. Two *GOES* satellites are operational at any time, both in a geostationary orbit.

geostationary orbit (geosynchronous orbit, Clarke orbit) A satellite orbit at a height of about 22,370 miles (36,000 km), in which the satellite travels in the same direction as the Earth's rotation, completing a single orbit in precisely one sidereal day, which means it remains permanently above the same point on the equator. Its field of view is almost an entire hemisphere. The writer Sir Arthur C. Clarke first suggested the possibility of placing satellites into geostationary orbit and the orbit is sometimes named after him.

geostrophic balance The condition above the boundary layer when the pressure-gradient force and Coriolis effect are equal and the wind blows parallel to the isobars.

geostrophic departure A difference between the wind speed that is observed and that of the geostrophic wind.

geostrophic equation The equation used to calculate the speed of the geostrophic wind: $V_g = (1/(2\Omega \sin \varphi\rho))(\delta p/\delta n)$, where V_g is the geostrophic wind velocity, Ω is the angular velocity of the Earth (= $15° \ hr^{-1} = 2\pi/24 \ rad \ hr^{-1} = 7.29 \times 10^{-5} \ rad \ s^{-1}$), φ is the latitude, ρ is the air density, and $\delta p/\delta n$ is the horizontal pressure gradient. $2\Omega \sin \varphi$ is known as the Coriolis parameter and sometimes designated by f, so the equation then becomes $V_g = (1/f\rho)(\delta p/\delta n)$.

geostrophic flux The transport of some substance or atmospheric property by the geostrophic wind.

geostrophic wind The wind above the boundary layer that blows almost parallel to the isobars, because the pressure-gradient force acting toward the center of low pressure is balanced by the deflection due to the Coriolis effect.

geostrophic-wind level (gradient wind level) The lowest height at which the wind flow becomes geostrophic. This is above the boundary layer, at an altitude of 1,650–3,300 feet (500–1,000 m).

geostrophic-wind scale A diagram from which the geostrophic wind speed can be read.

ghabi A wind that blows across the Mediterranean from the Sahara Desert to the Adriatic and Aegean Seas.

gharra Frequent, sudden, severe squalls, accompanied by heavy rain and thunderstorms, that cross Libya from the northeast.

ghibli A hot, dry wind of the sirocco type that blows in northern Libya.

glacial anticyclone (glacial high) The semipermanent region of high pressure that covers the Greenland ice sheet and Antarctica.

glacial period (ice age) A prolonged time during which a substantial part of the surface of the Earth is covered by ice.

glaciated cloud A cloud consisting wholly of ice crystals.

glacioisostasy The very slow rise in the level of the land that occurs after an ice sheet has melted.

glaze (clear ice) Clear, solid ice that covers surfaces with a layer up to one inch (2.5 cm) thick.

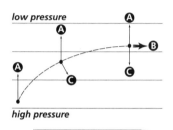

low pressure

high pressure

A *pressure gradient force*
B *wind direction*
C *Coriolis effect*

geostrophic wind

Global Area Coverage Oceans Pathfinder Project (GAC) A study of sea-surface temperatures using data from NOAA satellites.

Global Atmospheric Research Program (GARP) A project that ran from 1968 until the early 1980s to observe the atmosphere from space over a long period in order to estimate its variability and the processes occurring within it, with a view to improving the quality of weather forecasts.

Global Environmental Monitoring System (GEMS) A U.N. organization that was established in 1975 to acquire data pertaining to the natural environment and to make them available to governments and other organizations that need them.

Global Environment Facility (GEF) An organization established in 1990 and managed by the World Bank and United Nations Environment Program (UNEP) that provides practical assistance to the environmental improvement programs of governments.

Global Horizontal Sounding Technique (GHOST) A project forming part of the World Weather Watch that uses balloons floating at various constant-density levels for the direct sensing of the atmosphere.

Global Inventory Monitoring and Modeling Studies (GIMMS) A set of data on global vegetation held at the NASA Goddard Space Center.

Global Resource Information Database (GRID) An international organization that uses computers and software developed by NASA to analyze and integrate environmental information from different sources, including the World Health Organization and the Food and Agriculture Organization.

global warming The idea that the mean temperature of the atmosphere is increasing throughout the world.

global warming potential (GWP) The amount of climatic forcing exerted by a particular greenhouse gas. This is compared to the forcing exerted by carbon dioxide, which is given a value of 1. Water vapor, the gas with the strongest greenhouse effect, is not included, because the atmospheric content is highly variable and beyond our control.

gloom The condition in which daylight is markedly reduced by thick cloud or dense smoke, but horizontal visibility remains good.

glory An optical phenomenon in which a shadow cast onto a layer of cloud appears surrounded by one or more circles of light.

grab sampling A technique for obtaining a sample of air very quickly, so that the time taken to obtain the sample is insignificant when compared to the duration of the process or rate of change being studied.

gradient flow The horizontal movement of air when there is no friction and the isobars and streamlines coincide.

gradient wind The wind that flows parallel to the isobars at a speed resulting from the interplay of the pressure-gradient force, the Coriolis effect, and centripetal acceleration.

grape belt A stretch of land about 60 miles (96 km) along the southern shore of Lake Erie that benefits from the lake effect. Winters are less severe than in other parts of the United States at the same latitude.

grass minimum temperature The temperature registered by a minimum thermometer set in the open at the level of the tops of the blades of grass in short turf.

grass temperature The temperature registered by a thermometer set in the open with its bulb at the level of the tops of grass blades in short turf.

graupel (soft hail) Precipitation in the form of white, opaque ice pellets soft enough to flatten or smash into fragments when they hit a hard surface. The pellets are 0.1–0.2 inch (2–5 mm) in diameter, but can also be less than 0.04 inch (1 mm) across, and most are spherical.

gravitational force An attraction between bodies with a magnitude equal to the mass *(m)* of the body multiplied by the acceleration *(a)* it produces *(ma)*. This decreases with increasing distance between the bodies according to the inverse square law. The gravitational acceleration produced by the Earth is designated by *g* and has an average value of 9.8 m s^{-2} (32 ft s^{-2}).

gravitational settling The process by which solid particles fall from the air due to the force of gravity.

graybody A body that absorbs electromagnetic radiation uniformly at all wavelengths. The emittance *(E)* of a graybody is given by: $E = \varepsilon \sigma T^4$, where ε is the emissivity, σ is the Stefan–Boltzmann constant, and *T* is the absolute temperature. Many natural bodies are graybodies across a wide range of wavelengths.

green flash (green ray) A bright flash of emerald green light, lasting one to 10 seconds or sometimes longer, seen just above the horizon immediately after the Sun has set or immediately before it rises.

greenhouse effect Warming of the atmosphere due to the absorption and reradiation of heat by the molecules of certain gases (greenhouse gases). Radiation absorbed in the atmosphere makes it 54–72°F (30–40°C) warmer than it would be if it were transparent to radiation at all wavelengths. Many scientists believe human activities that release greenhouse gases may be increasing the natural greenhouse effect and that the resulting enhanced greenhouse effect will induce global warming.

greenhouse gas A gas that absorbs energy radiated from the surface or atmosphere of the Earth or another planet. The principal greenhouse gases are water vapor (H_2O), carbon dioxide (CO_2), nitrous oxide (N_2O), methane (CH_4), ozone (O_3), CFCs, and hydrofluorocarbons.

greenhouse period A time when there were no glaciers or ice sheets anywhere on Earth.

Greenland Icecore Project (GRIP) A drilling program sponsored by the European Science Foundation to retrieve an ice core 1.86 miles (3 km) long from Summit, the highest point on the Greenland ice sheet. On August 12, 1992, the drill reached bedrock at a depth of 9,938 feet (3,029 m). Ice at that depth is 200,000 years old. The core is stored, in sections, at the University of Copenhagen, Denmark.

Greenland ice sheet The layer of ice that covers 80% of the area of Greenland. The inland ice sheet is almost 1,490 miles (2,400 km) long in a north–south direction and 680 miles (1,100 km) wide at its widest point (77° N) with an area of 670,272 square miles (1,736,095 km²). Smaller ice caps and glaciers cover an additional 18,763 square miles (48,599 km²). Two north–south domes or ridges, one in the north and the other in the south, are where the ice is thickest. The southern dome is about 9,845 feet (3,000 m) thick at 63–65° N, and at about 72° N the northern dome is about 10,795 feet (3,290 m) thick.

Greenland Ice Sheet Project (GISP) A United States drilling program that retrieves ice cores from the Greenland ice sheet. The first core reached bedrock at a depth of about 9,843 feet (3,000 m) and a second core, GISP2, penetrated 5 feet (1.55 m) into bedrock, at a depth of 10,018.34 feet (3,053.44 m).

gregale A strong, northeasterly wind that blows across Malta and the adjacent parts of the Mediterranean region.

Grosswetterlage A classified catalog of large-scale airflow patterns over central Europe that is published daily by the central office of the German Weather Service (Deutscher Wetterdienst) at Offenbach, Germany.

ground fog Any fog that covers less than 60% of the sky.

ground frost Frost that forms when the air temperature is above freezing, but the temperature at ground level is below freezing.

ground heat flux (soil heat flux) The flow of heat into the ground by day and from the ground at night.

Groundhog Day February 2, the day on which an old tradition holds that the spring weather can be predicted. In parts of North America, it is the day when the groundhog is said to emerge from its hibernation. If it sees its shadow, it anticipates bad weather and returns to its hole for a further six weeks. If it cannot see its shadow, because the day is cloudy, it remains above ground, anticipating fine weather.

ground streamer A column of ionization that rises at the start of a lightning stroke from a point on the ground toward which a stepped leader is descending.

ground visibility Horizontal visibility measured at ground level or at the height of an airfield control tower.

ground water (phreatic water) Water below the ground surface in the zone of saturation (phreatic zone) where all spaces between soil particles are filled with water.

growing season That part of the year during which weather conditions permit the growth of agricultural and horticultural crops.

guba A strong squall that occurs at night along the southern coast of New Guinea during the summer monsoon.

gully squalls Violent squalls that blow over the coastal waters of the eastern South Pacific from mountain ravines in the Andes.

gust A sudden, sharp increase in wind speed that lasts for only a short time.

gust front (pressure jump line) A region immediately ahead of an advancing storm where unstable, warm air is being drawn into the storm cloud, producing strong, gusty winds.

gustiness factor A measure used to describe the intensity of the gusts generated by a particular wind.

gustnado A small tornado that forms in the gust front ahead of a cumulonimbus cloud containing a supercell.

haar A cold sea fog common along the eastern coast of Britain, especially in fine weather.

haboob A severe dust storm in northern Sudan, most often late in the day in summer.

Hadley cell Part of the general circulation of the atmosphere in which warm air rises close to the equator, losing most of its moisture as it does so, moves away from the equator at a high level, then descends close to the tropics. The subsiding air warms adiabatically, reaching the surface as warm, dry air. There are several Hadley cells in each hemisphere.

hail Precipitation in the form of hailstones.

hail day A day on which hail falls.

hail region One of 13 regions into which the United States is divided according to the frequency and intensity of the hail they experience.

hailshaft A column of falling hailstones visible beneath the cloud base.

hailstone An approximately spherical ice pellet that falls from a cumulonimbus cloud. Hailstones vary greatly in size, but most are 0.2–2 inches (5–50 mm) in diameter. Many hailstones consist of alternate layers of clear and opaque ice.

hailstreak A strip of ground completely covered by fallen hailstones.

hailswath An area of ground partly covered in hailstones.

hair hygrometer (hygroscope) An instrument that measures atmospheric humidity and gives a direct reading of the relative humidity. It exploits the fact that human hairs lengthen as the humidity increases and shrink as it falls.

halcyon days A period of calm, peaceful weather, especially in winter.

Hale cycle The approximately 22-year cycle during which the magnetic polarity of the Sun reverses. Climate records for the last 300 years show that severe droughts in the western United States occur twice in the course of the Hale cycle.

half-arc angle The angle bisecting the arc that extends from the point directly above the head of an observer to the horizon seen by the same observer.

halo A circle of white light surrounding the Sun or Moon, seen when the Sun or Moon appears as a white disk behind a layer of cirrostratus. The radius of a halo subtends an angle of 22°, or less commonly 46°, to the eye of the observer.

halogen A chemical element belonging to Group VII of the periodic table, i.e., fluorine, chlorine, bromine, iodine, or astatine.

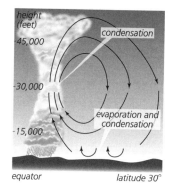

Hadley cell

halogen occultation experiment (HALOE) An experiment that measures the absorption of infrared radiation at sunrises and sunsets (occultations) and also concentrations of a range of gases, including ozone (O_3), nitric oxide (NO), nitrogen dioxide (NO_2), water vapor (H_2O), methane (CH_4), hydrochloric acid (HCl), and hydrogen fluoride (HF).

halon The commercial name for one of the bromofluorocarbons, which are chemical compounds used in fire extinguishers and implicated in the depletion of stratospheric ozone.

hard UV Ultraviolet radiation with a wavelength shorter than 280 nm.

harmattan A moderate or strong, hot, dry, dusty wind that blows only during the day across West Africa south of the Sahara.

haze A reduction in horizontal visibility, but not to below 1.2 miles (2 km), that occurs when aerosols absorb and scatter sunlight.

haze droplet A liquid droplet smaller than 1 μm (0.0004 inch) in diameter that forms by condensation onto a hygroscopic nucleus in air with a relative humidity (RH) greater than about 80% and remains suspended in the air. As the RH rises above 90%, haze droplets grow and begin to reduce visibility.

haze factor The ratio of the brightness of an object that is seen through haze or fog to the brightness of the same object seen through clear air.

haze horizon The top of a haze layer when seen from above and against a clear sky.

haze layer A band of haze extending to the surface, usually seen beneath a temperature inversion.

haze line The boundary between the upper surface of a haze layer and the clear air above it.

heat capacity (thermal capacity) The amount of heat energy that must be supplied to a substance in order to raise the temperature of that substance. It is measured in relation either to a unit mass of the substance, when it is known as the specific heat capacity (symbol c), or to a unit amount of the substance, when it is known as the molar heat capacity (symbol C_m).

heating degree days The number of degrees by which the mean daily temperature falls below a base level. In the United States the base level is 65°F (19°C). The value is used in calculations of fuel consumption for heating.

heat island An area that is markedly warmer than its surroundings and therefore resembles an island of warmth in a sea of cool air.

heat lightning Silent flashes of sheet lightning, most often seen on warm summer nights. The storm causing them is more than about 6 miles (10 km) away and all the sound waves have been either absorbed by the air or refracted upward by the effect of the decrease in air temperature with height.

heat stress, index of A numerical value describing the relationship between actual temperature, relative humidity, and apparent temperature, used to indicate whether people will feel comfortable under specified weather conditions.

heat thunderstorm A thunderstorm on a warm, humid afternoon, when the ground has been strongly heated, making the air above it unstable.

heat wave A period of at least one day, but more often several days or weeks, during which the weather is unusually hot for the time of year.

heavenly cross A sun pillar crossed by a horizontal bar.

heiligenschein An optical phenomenon in which the shadow of an observer on ground covered with vegetation has a bright, white light surrounding the head.

Heinrich event The relatively sudden calving of a large swarm of icebergs from North America into the North Atlantic. This is believed to have occurred six times between 70,000 and 16,000 years ago, during the most recent glaciation.

heliotropic wind A slight change in the wind direction that takes place during the course of the day due to the east-to-west progression of the surface area most intensely warmed by the Sun.

heterogeneous nucleation The freezing of supercooled water droplets onto freezing nuclei.

heterosphere That part of the atmosphere, above a height of about 50 miles (80 km), which has a heterogeneous (nonuniform) composition.

high An area in which the surface atmospheric pressure is higher than the pressure in the surrounding area.

high-altitude station A weather station located no lower than about 6,500 feet (2,000 m) above sea level, where the conditions it records are significantly different from those at sea level.

high cloud A cloud with a base at above 20,000 feet (6,000 m) altitude. Cirrus, cirrocumulus, and cirrostratus are high clouds.

high fog Fog on the upper slopes of a mountain that sometimes extends as stratus cloud over the valley on the lee side.

high föhn (stable-air föhn) The wind that develops when an upper layer of stable air is at a higher potential temperature than a lower layer. It occurs when cold, stable air moves against a mountain range. Stable air spills over the mountains and descends, warming adiabatically as it does so. It can also develop when subsiding air in an anticyclone is chilled by a cold land or sea surface.

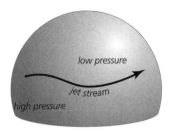

high index

high index A high value of the zonal index; the jet stream is strongly developed and blowing steadily in an approximately easterly direction.

high level inversion A temperature inversion 1,000 feet (300 m) or more above the surface, caused by the cooling of subsiding air in an anticyclone when the surface is much colder than the air.

high resolution Doppler interferometer (HRDI) An interferometer carried on the *Upper Atmosphere Research Satellite* that provides data on the temperature and winds throughout much of the stratosphere (but with a gap in the upper stratosphere) and most of the mesosphere.

high resolution picture transmission (HRPT) A method used by satellites to transmit images with a resolution of 0.7 mile (1.1 km), allowing them to depict features 0.5 mile (0.8 km) across.

hill fog (upslope fog) Fog that forms when moist air is made to rise, causing it to cool adiabatically.

historical analog model A climate model based on comparisons with weather conditions that obtained at some time in the past.

hoar frost Ice that forms as a thin layer of white crystals on grass, other herbs, shrubs, trees, spider webs, and other exposed surfaces.

homogeneous atmosphere A hypothetical atmosphere in which the air density remains constant at all heights.

homogeneous nucleation The spontaneous freezing of supercooled water droplets that occurs at very low temperatures in the absence of freezing nuclei.

homosphere That part of the atmosphere in which the chemical composition is homogenous (uniform). It extends from the surface to a height of about 50 miles (80 km) and comprises the troposphere and stratosphere.

hook pattern A distinctive shape that is often visible in radar images taken from directly above clouds associated with a supercell storm. It usually occurs at the edge of a mesocyclone and therefore indicates an immediate risk of tornadoes.

horse latitudes The latitudes of the subtropical highs, at approximately 30° in both hemispheres, where winds are light and variable or the air is calm, and sailing ships can be becalmed. Supplies of water sometimes ran low, and horses died and were thrown overboard.

hot belt The belt around the Earth within which the mean annual temperature exceeds 68°F (20°C).

hot lightning Lightning that starts forest fires. The current carried by the lightning stroke is sustained for a fraction of a second, generating enough heat to ignite dry material.

hot tower A narrow column of air that is rising rapidly by convection and is enclosed by a much larger volume of air that is rising very little or sinking.

humidification Deliberately adding moisture to the air, often with the aim of reducing the accumulation of static electricity.

humidifier A device that injects moisture into the air.

humidistat A device that monitors humidity. Humidistats are used to switch on dehumidifiers and humidifiers when the humidity departs from a selected value.

humidity The amount of water vapor present in the air. The term refers only to water that is present as a gas. Humidity can be measured as the mixing ratio, specific humidity, absolute humidity, and relative humidity. Relative humidity is the measure used in weather reports.

humidity coefficient A measure of the effectiveness for plant growth of the precipitation falling over a region during a specified period. It is calculated as $P/1.07^t$, where P is the amount of precipitation in centimeters and t is the mean temperature for the period in question in degrees Celsius.

humidity index A measure of the extent to which the amount of water available to plants exceeds the amount needed for healthy growth. It is calculated as $100W_s/PE$, where W_s is the water surplus and PE is the potential evapotranspiration.

humilis (hum) A species of cumulus clouds with flat bases and little vertical development. They represent cumulus that has failed to grow.

hurricane A tropical cyclone in the North Atlantic or Caribbean. The name is often applied to any tropical cyclone, regardless of where it occurs.

hurricane-force wind A wind of at least 75 mph (121 kmh⁻¹). This is force 12 on the Beaufort wind scale and category 1 on the Saffir/Simpson hurricane scale.

hurricane monitoring buoy A free-floating instrument package that detects the approach of a tropical cyclone and is designed to be expendable.

hydrodynamical equations A set of equations of continuity used in numerical forecasting that relate changes in the concentrations of the various constituents of the air to their transport, sources, and sinks.

hydrological drought A drought during which the ground water table falls markedly.

hydrologic cycle (water cycle) The constant circulation of water from the oceans to the atmosphere and back to the surface as precipitation.

hydrologic model A model used to simulate the behavior of hydrologic systems, such as drainage basins and river flow. It may be a small-scale physical device that mimics the real system, a computer simulation, or a sequence of mathematical calculations.

hydrometeor All water present in the atmosphere in the liquid or solid phase. The term includes every form of precipitation, including dew, frost, and virga (fallstreaks).

hydrometeorology The branch of meteorology specializing in the study of precipitation.

hydrosphere Water at or close to the surface of the Earth, including the oceans, icecaps, lakes, rivers, and ground water.

hydrostatic approximation The assumption that the atmosphere is in hydrostatic equilibrium.

hydrostatic equation An equation relating the weight of a column of air of unit cross-sectional area to the pressure exerted on it from above, at height p_2, and below, at height p_1. The weight of the air is equal to its mass *(M)* multiplied by its gravitational acceleration *(g)*. Then $Mg = p_1 - p_2$, provided the density of the air remains constant throughout the column and *g* is a constant. For a deeper column allowance must be made for the vertical pressure gradient. Then $\delta p/\delta z = -g\rho$, where *p* is pressure, *z* is height, ρ is density, and the minus sign indicates that density decreases with height.

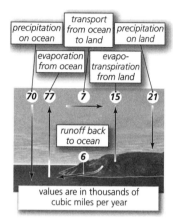

precipitation on ocean | transport from ocean to land | precipitation on land

evaporation from ocean | evapo-transpiration from land

70   77   7   15   21

runoff back to ocean

6

values are in thousands of cubic miles per year

hydrologic cycle

Gravitational acceleration decreases with height, but within the troposphere the effect is so small it may be ignored. It also changes with latitude. To allow for this, geopotential height is used rather than vertical distance.

hydrostatic equilibrium The condition of the atmosphere in which the weight of a parcel of air is exactly balanced by an upward pressure at its base.

hydroxyl A hydroxide molecule (OH).

hyetogram A chart that records the amount and duration of rainfall at a particular place.

hyetograph An instrument that measures rainfall and records it automatically as a line on a chart fixed to a rotating drum.

hyetography The study of the annual geographic distribution of precipitation and variations in it.

hygrogram A chart showing a continuous record of relative humidity over a period, usually of one week.

hygrograph A hygrometer that records changes in relative humidity as a line on a chart fixed to a rotating drum.

hygrometer An instrument that measures the humidity of the air.

hygroscopic moisture Water absorbed from the atmosphere by soil particles.

hygroscopic nucleus A cloud condensation nucleus made from a substance that absorbs water and swells in size as it does so, eventually dissolving into a concentrated solution.

hygrothermograph An instrument that provides a continuous record of both relative humidity and temperature.

hypsometer (hypsometric barometer) A barometer that calculates air pressure by measuring the temperature at which a liquid boils.

hythergraph A diagram comparing the climates of two or more places. Monthly mean temperature is plotted against mean precipitation and the points joined by straight lines, with the point for December linked to that for January, producing a closed figure. The shape and dimensions of the figure indicate the range of temperature and precipitation through the year, and the location of the figure indicates the overall temperature and precipitation.

hythergraph

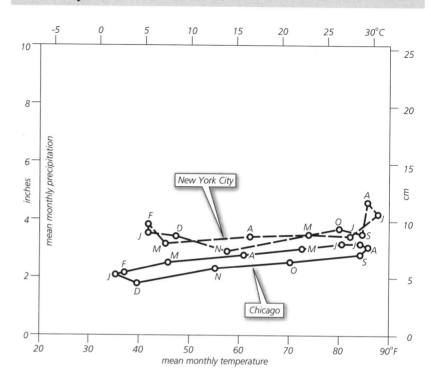

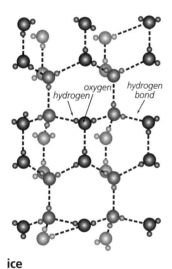

hydrogen   oxygen   hydrogen
bond

ice

ice   Water in the solid phase, comprising water molecules arranged in a regular, repeating pattern (a lattice). Their movement is restricted to vibrating about a fixed point. Each molecule is linked by hydrogen bonds to three neighboring molecules, producing a very open structure.

ice blink   A white gleam visible above the horizon when pack ice is present in the distance. It is caused by the reflection of light from the ice.

icebow   An arc of white light resembling a rainbow and formed by the reflection and refraction of light through ice crystals.

ice concentration   The percentage of an area of ocean surface covered by ice.

ice core   A sample of ice for scientific study obtained by drilling vertically into a glacier and extracting a long cylinder.

ice-crystal cloud   Cloud consisting entirely of ice crystals.

ice-crystal haze   Haze consisting entirely of ice crystals.

ice crystals   The form in which molecules arrange themselves when water freezes. Ice crystals are classified in ten types: plates (flat, hexagonal rings); stellars (rings with six points); columns (cylinders, hexagonal in cross-section and sometimes joined together); capped columns (columns that have a bar at each end, sometimes joined together); needles (resembling splinters, sometimes joined); and spatial dendrites (crystals with many fine branches). Irregular crystals are clumped together and have no regular shape. There are three more categories for graupel, sleet, and hail. A standard symbol is used for each type.

ice day A day when the air temperature does not rise above freezing and when ice on the surface of water does not thaw.

ice evaporation level The height at which ice crystals entering dry air will change directly into water vapor by sublimation.

ice fog Fog that forms when relatively warm water is suddenly released into very cold air and water vapor changes directly to ice crystals.

icehouse period A period when glaciers and ice sheets reached their maximum extent.

Icelandic low One of the two semipermanent areas of low pressure in the Northern Hemisphere (the other is the Aleutian low), centered between Iceland and Greenland, between about 60° N and 65° N.

ice nucleus Any small particle onto which supercooled water will freeze.

ice pellets Precipitation in the form of ice particles that are transparent or translucent, spherical or irregular in shape, and less than 0.2 inch (5 mm) in diameter.

ice period The time that elapses between the first fall of snow in winter and the melting of the last patches of snow in spring.

ice prisms (diamond dust) Ice crystals so tiny they are barely visible as they hang suspended in the air, twinkling as they catch and reflect the sunlight.

ice splinter A fragment of ice that becomes detached from an ice crystal as the crystal grows inside a cloud.

ice storm A storm of freezing rain that deposits thick layers of ice onto structures such as radio masts, the rigging of ships, trees, and telephone wires.

icing The formation of ice on the surfaces of the leading edges of the wings, tailplanes, and tail fins of aircraft.

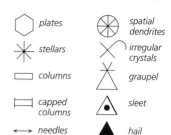

ice crystal (standard symbols)

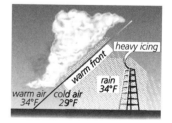

ice storm

icing level The lowest altitude at which an aircraft is expected to experience icing in a particular locality under the prevailing weather conditions.

ideal gas A gas composed of very small molecules in which there are no forces acting between the molecules. Such a gas obeys the equation of state under all conditions.

immission The receipt of a substance, such as an atmospheric pollutant, from a distant source; the opposite of emission.

impaction The removal of solid particles from the air through their collision with and adherence to surfaces.

impingement The bringing of two or more substances or objects into contact with each other.

improved stratospheric and mesospheric sounder (ISAMS) An instrument carried on the *Upper Atmosphere Research Satellite* that measures the temperature in the stratosphere and mesosphere and the atmospheric concentrations of ozone (O_3), nitric oxide (NO), nitrogen dioxide (NO_2), nitrous oxide (N_2O), nitric acid (HNO_3), dinitrogen pentoxide (N_2O_5), water vapor (H_2O), methane (CH_4), and carbon monoxide (CO).

inactive front (passive front) A front or part of a front that has very little cloud or precipitation associated with it.

inclination The angle between the orbital plane of a body and a reference plane centered on the body about which it is orbiting (in the case of the Earth, the plane of the ecliptic).

inclination of the wind The angle between the wind direction and the isobars.

inclined orbit An orbit between the geostationary (inclination 0°) and polar (inclination 90°) orbits.

incus (anvil) A supplementary cloud feature comprising the cirriform cloud at the top of a cumulonimbus cloud. It is often swept into the shape of an anvil by the wind.

indefinite ceiling The condition in which vertical visibility cannot be measured precisely, because it is determined by fog, haze, blowing snow, sand, or dust.

index cycle A progressive change in the zonal index that typically lasts three to eight weeks. The zonal index represents the difference in pressure between two latitudes, usually 33° N and 55° N. When the cycle commences, the polar front and its jet stream are aligned

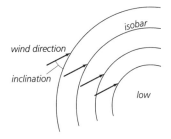

inclination of the wind

approximately from west to east, with a small number of Rossby waves along them. During an index cycle, these waves become increasingly extreme until the pattern breaks down altogether. Then the flow of air on either side joins, forming isolated cells. Finally, the original flow reestablishes itself.

Indian Ocean Experiment (INDOEX) An international field experiment using data from Meteosat-5 that studies air pollution, clouds, interactions between clouds, and solar radiation in the Intertropical Convergence zone over the Indian Ocean.

Indian summer A period of warm weather with clear skies that occurs in late September and October in the northeastern United States and especially in New England.

indirect cell Ferrel cells, which form part of the general circulation of the atmosphere driven not by convection, but by the direct cells to the north and south of them.

industrial climatology The application of climatological studies to industry in order to determine the influence of climate on a particular operation.

industrial meteorology The study of ways in which day-to-day weather conditions affect a particular industry.

inertial reference frame A location that is not accelerating and from which the acceleration of other bodies can be observed accurately.

inferior image An image produced by a mirage below the position of the object that is its source.

infrared radiation Electromagnetic radiation with a wavelength from 0.7 μm to 1 mm.

initial condition A value used as the base from which later values are calculated.

Initial Joint Polar System (IJPS) A system of polar-orbiting meteorological satellites that will come into operation in 2003, operated jointly by the National Oceanic and Atmospheric Administration and Eumetsat, to enhance and supplement climate monitoring and numerical forecasting.

insolation The amount of solar radiation that reaches the surface over a unit area of the surface of the Earth.

instability The tendency of air to continue to rise once it has begun to do so.

instability line Any line along which the air is subject to vigorous convection, and is therefore unstable, that is not associated with a frontal system.

insular climate The climate of an oceanic island or coastal region, where the influence of the ocean is greater than that of the nearest large land mass.

intensification An increase in the pressure gradient that takes place over hours or days, leading to a strengthening of the winds.

interception The catching of precipitation by plant or other surfaces, or of sunshine by objects that cast shadows.

interference The effect of imposing one set of waves upon another, producing a new wave with a different form and characteristics from either of the waves from which it is made.

interferometer An instrument used to measure the wavelength of radiation.

interglacial A prolonged period of warmer climates separating two glacial periods.

Intergovernmental Panel on Climate Change (IPCC) An organization established in 1988 by the World Meteorological Organization and the United Nations Environment Program to assess the scientific, technical, and socioeconomic information relevant to changes in climate caused by human activity and their consequences.

International Cloud Atlas A book published by the World Meteorological Organization containing the pictures and definitions against which clouds are classified. The *Atlas* is in two volumes; there is also a single-volume, abridged version.

International Cloud Code A system, adopted in 1929, that divides clouds into four types: high, middle, low, and clouds with vertical development.

international index numbers A system devised by the World Meteorological Organization that identifies each meteorological observing station by a number.

International Phenological Gardens (IPG) A network of European gardens, all of which grow genetically identical clones of trees and shrubs and keep records of the dates of phenological events, such as the unfolding of leaves, growth of shoots, flowering, leaf coloring, and leaf fall.

International Polar Year The year from August 1882 until August 1883, during which scientists collaborated internationally for the first time to investigate the arctic environment, and especially arctic meteorology, auroras, and the Earth's magnetic field.

International Satellite Cloud Climatology Project (ISCCP) A program that began collecting data on July 1, 1983. The data are analyzed to obtain information about the location, extent, and types of cloud over the entire surface of the Earth.

International Satellite Land Surface Climatology Project (ISLSCP) A project in which satellite images are used to map vegetation.

international synoptic code A code approved by the World Meteorological Organization for the transmission of meteorological data in which each element is encoded as a series of five-digit numerals.

interpluvial A period when the climate was drier than in the pluvial periods prior to or succeeding it.

interstade (interstadial) A time of warmer weather, lasting 1,000–2,000 years, during a glacial period.

intertropical convergence zone (ITCZ) A belt surrounding the Earth and close to its equator where the trade winds of the Northern and Southern Hemispheres meet and convergence causes the air to rise.

intertropical front The name formerly given to the intertropical convergence zone.

intortus A variety of cirrus clouds consisting of filaments that curve irregularly and appear to be entangled haphazardly.

inverse square law A law stating that the magnitude of a physical quantity is inversely proportional to the square of the distance between the source of that quantity and the place where it is experienced.

inversion The condition in which the air temperature increases with height, rather than decreasing.

invisible drought A drought in which precipitation falls, but the amount is insufficient to recharge aquifers when the water lost by evaporation and transpiration is deducted. Consequently, river levels and water tables remain low and plants continue to suffer stress.

ionosphere A layer of the atmosphere in which photoionization causes the separation of positively-charged ions and negatively-charged electrons. The thickness of the ionosphere and density of ions within it vary daily, seasonally, and with latitude. Generally, the ionosphere extends from a height of about 37 miles (60 km) to about 620 miles

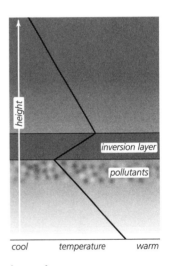

inversion

(1,000 km), although the ion density is greatest from about 50 miles (80 km) to 250 miles (400 km).

iridescence Shimmering colors produced by the diffraction of sunlight or moonlight by oil molecules, small water droplets, or ice crystals of approximately uniform size.

iridescent clouds Clouds that are partly brightly colored, most often with patches of red and green but sometimes with violet, blue, or yellow.

irisation The appearance of colors around the edges of clouds due to the diffraction of sunlight by supercooled water droplets or ice crystals.

irradiance The rate at which radiant solar energy flows through a unit area perpendicular to the radiation beam.

irregular crystal A particle of snow consisting of very small crystals that have formed erratically, so the resulting particle has an irregular shape.

isallobar A line drawn on a chart to link points at which the atmospheric pressure rose or fell by the same amount over the same specified time.

isallobaric An adjective describing a constant or equal change in atmospheric pressure over a spatial distance or a specified time.

isallotherm A line drawn on a chart to link points at which the atmospheric temperature rose or fell by the same amount over the same specified time.

isanomolous line (isanomal) A line drawn on a chart to link points where the value of a particular meteorological quantity varies from the regional average by the same amount, i.e., to join points of equal anomaly.

isentrope A line joining points of equal potential temperature.

isentropic analysis The extraction from radiosonde data of data on winds, pressures, temperatures, and humidities across several isentropic surfaces.

isentropic chart A synoptic chart on which meteorological elements, such as pressure, temperature, humidity, and wind, are plotted on an isentropic surface.

isentropic mixing Mixing of air that takes place across an isentropic surface.

isentropic surface A surface over which entropy is everywhere the same.

isentropic thickness chart A chart showing the thickness of an atmospheric layer bounded above and below by isentropic surfaces.

isentropic weight chart A chart showing the difference in pressure between two isentropic surfaces.

isobar A line drawn on a map to join points where the atmospheric pressure is the same.

isobaric map A map showing the distribution of atmospheric pressure at any given height above sea level.

isobront A line drawn on a map to link places at which thunderstorms reached the same stage of development at the same time.

isodrosotherm A line drawn on a synoptic chart to link places where the dewpoint temperatures are the same.

isogradient A line drawn on a map to link places where the horizontal gradient of pressure or temperature is the same.

isohel A line drawn on a map that joins places that experience equal numbers of hours of sunshine.

isohume A line drawn on a map to link places of equal humidity.

isohyet A line drawn on a map that joins places receiving the same amounts of rainfall.

isokeraun (isoceraun) A line drawn on a map to link places that experience the same frequency or intensity of thunderstorms.

isoneph A line drawn on a map to link places that are equally cloudy.

isonif A line drawn on a map to link places that received equal amounts of snowfall.

isopectic A line drawn on a map to link places where winter ice begins to form at the same time.

isophene A line drawn on a map to connect places at which a particular stage in plant development occurred on the same date.

isopleth A line or surface shown on a map or chart that connects points that are equal in respect of some quantity.

isoprene A volatile hydrocarbon compound (C_5H_8) synthesized by plants and emitted by plants, especially deciduous trees. Isoprenes and terpenes (made from isoprene molecules) play an important part in the formation of ozone. In urban areas isoprene makes an important contribution to the formation of photochemical smog; in rural areas it is the predominant hydrocarbon involved.

isopycnal A line or surface that is drawn on a map or chart to connect points of equal air density.

isoryme A line drawn on a map to link places where the incidence of frost is the same.

isosteric surface A surface across which air density remains constant.

isotach A line drawn on a map to join places that experience winds of the same speed.

isothere A line drawn on a map to link places where average summer temperatures are the same.

isotherm A line drawn on a map or tephigram to join points that are at the same temperature.

isothermal equilibrium (conductive equilibrium) The condition of a large body of air that is at the same temperature throughout.

isothermal layer The lower part of the stratosphere, in which the temperature remains constant with height.

isotropic Having properties that change independently of direction.

January thaw (January spring) A period of mild weather that sometimes occurs in late January in parts of the northeastern United States and in Britain.

jauk A föhn wind in the Klagenfurt Basin, Austria.

jet-effect wind A wind that is accelerated by funneling.

jet streak A region within the jet stream where the wind speed is higher than it is elsewhere.

jet stream A winding ribbon of strong wind in the upper troposphere or lower stratosphere. Typically, it is thousands of miles long, hundreds of miles wide, and several miles deep. There are several jet streams. The polar front and subtropical jet streams blow from west to east in both hemispheres. In summer the easterly jet blows from east to west across Asia, southern Arabia, and into northeastern Africa.

Jevons effect The effect on the measurement of rainfall caused by the rain gauge itself, which disturbs the flow of air past it. The effect is too small to be of importance.

Joseph effect The tendency for a particular type of weather to persist and repeat itself, known as the Joseph effect because of the dream Pharaoh described to Joseph: "Behold, there come seven years of great plenty throughout all the land of Egypt: And there shall arise after them seven years of famine" (Genesis 41. 29–30).

junk wind The name for the southerly or southeasterly monsoon wind in Thailand, Vietnam, China, and Japan. The wind is favorable to the sailors of junks (traditional flat-bottomed sailing ships).

junta A mountain-gap wind that blows through passes in the Andes.

juran (joran) A cold northwesterly wind in the Jura Mountains, near Geneva, Switzerland.

kachchan A hot, dry, westerly or southwesterly föhn wind that blows from the mountains of central Sri Lanka in June and July, during the southwest monsoon.

kal baisakhi A short-lived squall at the start of the southwest monsoon in Bengal.

kamikaze A "divine wind" (in Japanese); it was a tropical cyclone that saved Japan from a Mongolian invasion in 1281 by destroying the ships carrying the invading forces.

karaburan A hot, dry, east-northeastly wind of up to gale force that blows across the Central Asian deserts from early spring to the end of summer.

karema A strong easterly wind that blows across Lake Tanganyika, East Africa.

karif (kharif) A hot, dusty, southwesterly wind, often reaching gale force, that blows across the coast of Somalia on the southern side of the Gulf of Aden during the southwest monsoon.

Karman vortex street A series of vortices that develops when a fast-moving flow of air passes an obstruction, such as a building.

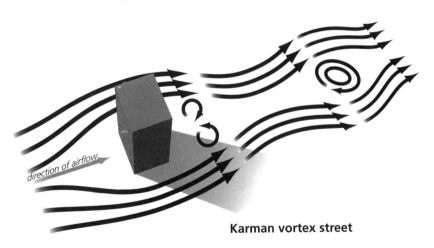

direction of airflow

Karman vortex street

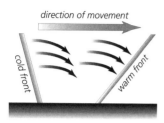

direction of movement

kata-front

katabaric (katallobaric) An adjective describing a phenomenon associated with a fall in atmospheric pressure.

katabatic wind (drainage wind, fall wind, gravity wind) A cold wind that blows downhill across sloping ground.

kata-front A front at which the air in the warm sector is subsiding relative to the cold air on either side of the front.

kaus (cowshee, sharki) A southeasterly wind that blows over the Persian Gulf ahead of a depression.

Kelvin waves Ocean waves associated with El Niño events that take about two months to cross the equatorial Pacific. Their amplitude is measured in tens of meters at the thermocline and their wavelength in thousands of meters.

Kew barometer A mercury barometer that allows the air pressure to be read directly from a scale at the top of its tube and requires no adjustment to the level of mercury in its reservoir.

khamsin A hot, dry, southeastly wind in Egypt and the Sudan.

killing frost A drop in temperature that kills plants or prevents them from reproducing.

kinematic viscosity The coefficient of viscosity of a fluid divided by its density.

kinetic energy The energy of motion, usually defined in terms of the amount of work a moving body could do if it were brought to rest. It is equal to $mv^2 \div 2$, where m is the mass of the moving body and v is its speed. For a rotating body, it is equal to $I\Omega^2 \div 2$, where Ω is the angular velocity and I is the moment of inertia.

Kirchhoff's law A law stating that the amount of radiation of a particular wavelength emitted by a body is equal to the amount absorbed by that body at the same wavelength. This is expressed as $\varepsilon\lambda = \alpha\lambda f(\lambda,T)$, where $\varepsilon\lambda$ is the emissivity at a stated wavelength, $\alpha\lambda$ is the absorptivity at the same wavelength, and $f(\lambda,T)$ is a function that varies with the wavelength (λ) and surface temperature (T).

kloof wind A cold, southwesterly wind across Simons Bay, South Africa.

knik wind A strong, southeasterly wind around Palmer, Alaska.

knot (kt) A unit of speed of one nautical mile per hour. By international agreement, 1 kt = 1.15 mph = 1.852 kmh^{-1}. The United States adopted the international knot in 1954, but British ships and aircraft continue to use a knot equal to 1.00064 international knots.

koembang A dry, warm, southerly or southeasterly wind of the föhn type that blows in Java, Indonesia.

kona A southwesterly wind that blows across Hawaii about five times a year.

kona cyclone (kona storm) A persistent, slow-moving, subtropical cyclone that delivers most of the winter rainfall in Hawaii.

Köppen climate classification A system for categorizing climates generically according to their temperatures and aridity. Köppen's is the climate classification scheme most widely used by geographers.

kosava A strong, usually cold, easterly or southeasterly wind that blows through the Iron Gate gorge on the Danube River, on the border between Romania and Serbia. It continues through the Carpathian Mountains and onto the Hungarian plain.

Krakatau (Krakatoa) A volcanic island in the Sunda Strait between Java and Sumatra, Indonesia, that erupted violently in 1883, ejecting about 5 cubic miles (21 km^3) of solid particles to a height of more than 19 miles (30 km). During the three years it took for the dust to settle there was a 10% decrease in the intensity of solar radiation and a small but significant fall in average temperatures.

krypton (Kr) A colorless gas that is present in the air in trace amounts. It is a noble gas with atomic number 36, relative atomic mass 83.80, density (at sea-level pressure and 32°F [0°C]) 0.0000021 ounces per cubic inch (0.0000037 g cm^{-3}). It melts at –250°F (–156.6°C) and boils at –242.1°F (–152.3°C).

Kyoto Protocol An international agreement drawn up in 1997 to provide guidelines for the implementation of the United Nations Framework Convention on Climate Change.

Labor Day storm A hurricane that struck southern Florida on Labor Day, in 1935.

lacunosus A variety of clouds that appear as patches, layers, or sheets with approximately round holes distributed more or less evenly so the cloud and holes form a pattern reminiscent of a net.

lake breeze A wind produced when air over land rises by convection and is replaced by cool, moist air drawn from over the surface of a lake.

lake effect A modification in the characteristics of air as it crosses a large expanse of water that is entirely enclosed by land. Temperatures are more moderate and the air more humid on the lee side of the lake and the precipitation is greater over high ground.

lake-effect snow Snow that falls in winter on the lee side of a large lake entirely enclosed by a large land mass. Cold, dry air is warmed by contact with the water and large amounts of water evaporate into it. When it reaches the other side it is chilled, the water vapor condenses, and there are heavy falls of snow.

Lambert's law A law stating that the intensity of the radiation emitted in any direction from a unit surface area of a radiating body varies according to the cosine of the angle between the direction of the radiation and a line perpendicular to the radiating surface.

lambing storm In parts of Britain, a light fall of snow in the spring, when ewes are lambing.

Lamb's classification A catalog of the type of weather conditions experienced over the British Isles every day from January 1, 1861, until February 3, 1997. The catalog describes the direction the weather systems were moving (N, NE, E, SE, S, SW, W, and NW), and classifies the systems as cyclonic, anticyclonic, or unclassifiable.

Lamb's dust veil index (dust veil index, DVI) A set of numerical values that quantify the climatic effect of dust ejected into the atmosphere by a volcanic eruption.

laminar boundary layer The layer of air in direct contact with a surface, in which air molecules are able to move only slowly, due to viscosity. Flow in this layer is laminar and parallel to the surface.

laminar flow The motion of a fluid that occurs smoothly, the fluid forming sheets that lie along streamlines parallel to each other.

land and sea breezes Winds that blow in coastal regions and along the shores of large lakes. A sea or lake breeze usually begins in late morning and reaches a maximum in the middle of the afternoon. At night the land breeze blows from the land toward the sea.

land blink A yellow glow seen in polar regions over an extensive snow-covered surface.

land contamination In satellite imaging, the effect of the footprint, about 30 miles (50 km) diameter, that occurs near coasts due to the mixing of data from land and sea.

Landsat A U.S. satellite program for the remote sensing of the Earth's resources, including studies of weather systems and climates. The

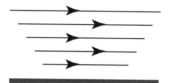

laminar flow

sea breeze day

land breeze night

land and sea breezes

first satellite was launched in 1972, called the *Earth Resources Technology Satellite 1 (ERTS 1)*. The name of the program and the satellites was changed to Landsat in 1975.

land sky The dark color of the underside of a cloud seen in polar regions over a land surface that is free from snow.

landspout (nonsupercell tornado) A tornado that develops in a relatively weak cumulonimbus cloud containing no mesocyclone or supercell.

langley (ly) A unit of solar radiation equal to one calorie per square centimeter per minute. The solar constant is equal to 1.98 langleys.

La Niña A strengthening of the southeasterly winds in the equatorial South Pacific that occurs at intervals of two to seven years. It is the opposite of an E1 Niño and completes a full ENSO event.

lapse line A curve on a graph showing the change of temperature with height in the free atmosphere.

lapse rates The rates at which a rising parcel of air cools with increasing height. The dry adiabatic lapse rate (DALR) is equal to 5.38°F per thousand feet (9.8°C km^{-1}). When the parcel of air reaches the lifting condensation level, the water vapor it carries will start to condense into liquid droplets. The air then cools at the saturated adiabatic lapse rate (SALR) of about 2.75°F per thousand feet (5°C km^{-1}).

large ion A charged particle of dust or other substance that exists as an aerosol, usually consisting of an ion attached to an Aitken nucleus.

laser altimeter An altimeter that calculates height above the surface by transmitting a laser beam directly downward and measuring the time that elapses between the transmission and the arrival of its reflection.

last glacial maximum The time, about 21,000 years ago, when the last ice age reached its greatest intensity. The ice sheets were then at their greatest extent, the global mean temperature was about 9°F (5°C) cooler than today, and climates everywhere were drier.

late-glacial The period when the most recent ice age was drawing to its close, from about 14,000 years ago until about 10,000 years ago.

latent heat The energy that is absorbed or released when a substance changes its phase between solid and liquid, liquid and gas, and directly between solid and gas. The latent heat of freezing and melting of water at 32°F (0°C) is 334 joules per gram (80 cal g^{-1}). The latent heat of condensation and vaporization is 2,501 J g^{-1} (600 cal g^{-1}) at

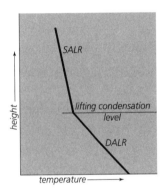

lapse rates

32°F (0°C). The latent heat of deposition and sublimation is 2,835 J g^{-1} (680 cal g^{-1}) at 32°F (0°C).

latent instability The condition in which a parcel of air that acquires sufficient kinetic energy to rise through a layer of stable air becomes unstable once it is above the level of free convection.

lateral acceleration The acceleration of air in a direction perpendicular to the wind direction.

laws of motion The three laws proposed by Sir Isaac Newton (1642–1727) to describe the way in which bodies respond to the forces acting on them. (1) A body will continue in a state of rest or uniform motion unless an external force acts on it. (2) The momentum of a body changes at a rate proportional to and in the same direction as the force acting on it. (3) If one body exerts a force upon another body, the second body exerts an equal force in the opposite direction upon the first body.

layer cloud A stratiform cloud of limited vertical extent that resembles a sheet.

leading-edge effect As air moves from one surface to another, only the air in immediate contact with the surface is affected by the new conditions and the altered boundary layer spreads downwind with only its lower part (the leading edge) fully adjusted to the new conditions.

lee The adjective describing the side of an obstacle sheltered from the wind. A lee shore is the shore on the lee side of a ship, so the wind tends to drive a ship toward the lee shore.

lee depression A depression that forms in a westerly airflow on the lee side of a range of mountains, which is aligned north and south.

lee trough (dynamic trough) A trough of low pressure that forms on the lee side of a long, north–south mountain barrier lying across a west–east airflow.

lee waves (standing waves) Waves that develop in stable air on the lee side of a mountain.

lenticular cloud A type of altocumulus lenticularis (Ac_{len}) often seen on the lee side of a mountain as a series of lens-shaped clouds (sometimes called wave clouds) extending downwind.

lenticularis (len) A species of clouds most commonly found in association with cirrocumulus, altocumulus, and stratocumulus. The cloud has the shape of a lens, usually with well-defined outlines, and sometimes has a very dramatic appearance. It can resemble a flying saucer!

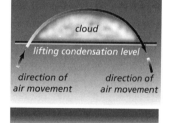

cloud

lifting condensation level

direction of air movement   *direction of air movement*

lenticularis

leste A hot, dry easterly or southeasterly wind blowing from Africa to Madeira and the Canary Islands.

levanter (solano, levante, llevante) A strong easterly wind, sometimes of gale force, that blows across southern Spain and the Strait of Gibraltar.

levantera A persistent easterly wind over the Adriatic Sea that usually brings cloudy weather.

leveche A hot, dry wind of the sirocco type that blows from the south over southeastern Spain.

level of free convection The height at which a parcel of air that is forced to rise through a conditionally unstable atmosphere changes from being cooler than the surrounding air to being warmer than it.

level of nondivergence The height at which there is no convergence or divergence of air in a column of air that is rising or subsiding.

libeccio A southwesterly wind that blows across Corsica and Italy.

lifting The forced ascent of air as it crosses high ground, is undercut by denser air at a front, or where airstreams converge.

lifting condensation level The altitude at which rising air cools to the dewpoint temperature and water vapor starts to condense into droplets.

light air In the Beaufort wind scale, force 1, which is a wind blowing at 1–3 mph (1.6–5 kmh^{-1}).

light breeze In the Beaufort wind scale, force 2, which is a wind blowing at 4–7 mph (6.4–11.3 kmh^{-1}).

lightning An electrical discharge that neutralizes a charge separation, which has accumulated within a cloud, between two clouds, or between a cloud and the ground.

lightning channel The path, about 8 inches (20 cm) across, along which a lightning stroke travels.

lightning conductor A metal rod that projects upward beyond the highest point of a building or other structure and is connected to the ground by a metal strip or cable with less than 10 ohms resistance. Lightning discharges are attracted to the metal, which guides them to the ground where they are earthed. A cone with an apex angle of 45° around the metal rod is protected from lightning strike.

lightning stroke The stepped leader, return stroke, and dart leader, which together comprise a flash of lightning.

linear acceleration Increase in speed in a straight line. It occurs in the entrance region of a jet stream and wherever air is accelerated in the direction of a wind.

Linke scale A set of cards, comprising only the even numbers from 2 to 26, used to measure the blueness of the sky. Each card is a different shade of blue and the observer uses an odd number if the sky shade falls between those of two cards.

liquid The phase of matter between the solid and gaseous phases, in which atoms or molecules are joined into small groups that move freely in relation to each other.

lithometeor Any solid particle suspended in or transported by the air.

Little Ice Age A period of cold weather that began in the 16th century and lasted until the early 20th.

llebetjado A hot, gusty wind that blows from the Pyrenees across Catalonia, Spain.

llevantades A levanter wind that brings especially stormy weather.

local climate The climates within an area with a distinct type of surface, e.g., of a forest, farm, or city.

local extra observation A weather observation taken at an airport at frequent intervals, often of 15 minutes, when conditions are close to the minima, for visibility, cloud base, etc., specified for taking off and landing, and flight operations are imminent.

local forecast A weather forecast for a small area for up to about two days ahead.

local storm A storm affecting only a small area.

local wind A wind that differs, in direction or strength, from the winds associated with the general distribution of pressure. Many local winds have their own names.

lofting A pattern made by a chimney plume that widens with increasing distance from the smokestack and rises gently as it disperses.

logarithmic wind profile The variation of wind velocity with height throughout the planetary boundary layer above the laminar boundary layer. This is expressed by $\bar{u}_z = (u*/k) \ln (z/z_0)$, where $\bar{u}_z$ is the mean wind speed in meters per second at a height $z, u*$ is the friction velocity, k is the von Karman constant, and z_0 is the roughness length in meters (ln means "natural logarithm").

lofting

London smog incidents Two episodes in London, England, each lasting several days, during which very wet smog was trapped beneath an inversion. In the first episode, in December 1952, more than 4,000 people in the Greater London area died as a direct consequence of the smog; about 700 people died in the second smog incident in December 1962.

long-range forecast A weather forecast for a period up to two weeks and sometimes up to one month ahead.

loom A glow of light just below the horizon caused by the refraction of light passing from cooler air aloft to warmer air below.

looming An adjective describing an object on the surface that appears to be raised above the surface as a superior image produced by a mirage.

looping The pattern made by a chimney plume that descends toward the ground and then rises again, repeating this until the gases and particles have dispersed.

looping

low An area of low atmospheric pressure.

low cloud A cloud with a base below 6,500 feet (2,000 m). Stratus, stratocumulus, and nimbostratus are low clouds.

lower atmosphere The troposphere, including the planetary boundary layer.

lowering The emergence of a mass of cloud beneath the base of a large cumulonimbus cloud, lowering the cloud base.

low index A low value of the zonal index, which means the Rossby waves in the jet stream and polar front are well developed, so the jet stream follows a sinuous path.

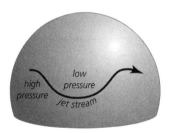

low index

lull A temporary fall in the wind speed or cessation of precipitation.

luminous meteor Any atmospheric phenomenon that appears as a pattern of light in the sky.

lysimeter An instrument used to measure the rate of evapotranspiration.

mackerel sky Cirrocumulus cloud in which the individual units are swept into long, parallel rows, forming a pattern reminiscent of the scales on the back of a mackerel fish.

macroburst A strong downdraft from the base of a cumulonimbus cloud.

macroclimate The climate typical of a very large area, such as a continent or the entire Earth.

macroclimatology The study of macroclimates.

macrometeorology The study of the atmosphere at the largest scale, including the general circulation and the development and behavior of air masses and large weather systems.

Madden-Julian Oscillation (MJO) An atmospheric disturbance that begins over the Indian Ocean and travels eastward as a wave with a 30- to 60-day period, causing a warming in the lower atmosphere; several MJO cycles can amplify the effect.

maestrale A cold northerly or northwesterly wind that blows over northern Italy.

maestro A northwesterly wind that blows across Italy, the Adriatic Sea, and the western shores of Sardinia and Corsica, bringing fine weather.

magnetic declination The difference between the direction of the north and south magnetic poles and the north and south geographic poles.

magnetic wind direction The wind direction measured by a magnetic compass.

magnetopause The boundary between the magnetosphere and the bow shock wave where the solar wind interacts with the plasma of the magnetosphere.

magnetosheath A region of the magnetosphere between the bow shock wave and the magnetopause.

magnetosphere The region of space above the exosphere, consisting of a plasma that is constantly maintained through bombardment by the solar wind.

mai-u (plum rains) Very heavy rains that fall during the first half of June along the valley of the Chang Jiang (Yangtze River), China.

maize rains Heavy and prolonged rains that fall in East Africa between February and May.

mammatus A supplementary cloud feature on the underside of a large anvil that appears as many smooth, udder-shaped protrusions from the cloud base.

mango showers (blossom showers) Rain showers, produced by occasional thunderstorms, that fall in southern India in April and May.

mares' tails Cirrus fibratus cloud appearing as long, wispy strands that curl at the ends.

marin A southeasterly wind of the sirocco type in the land bordering the Gulf of Lions, France.

marine forecast A weather forecast prepared for the crews of ships at sea.

marine meteorology The branch of meteorology specializing in the study of weather conditions over the open ocean, coastal seas, coastal land areas, and islands.

maritime air Air that forms air masses over all of the oceans. It is moist and its temperature is less extreme than that of continental air forming in the same latitude.

maritime climate (oceanic climate) A climate produced by maritime air. The seasonal temperature range is smaller than that of an area with a continental climate and precipitation is higher.

mass flux The rate at which the volume of a given mass of air changes.

mathematical climate An ancient Greek system that classified climates according to the height of the Sun above the horizon. The world was divided into three latitudinal belts: the "winterless" (later Torrid) Zone; "summerless" (later Frigid) Zones; and "intermediate" (later Temperate) Zones. These were separated by the Arctic and Antarctic Circles and the tropics of Cancer and Capricorn. Later, in the 4th century BCE, Aristotle calculated the latitudes defining the belts.

Maunder minimum The period from 1645 to 1715 when very few sunspots were observed. This coincided with the coldest part of the Little Ice Age.

maximum thermometer A mercury thermometer that records the highest temperature reached since it was last reset. As the temperature rises, the force of expansion pushes mercury past a constriction in the tube, but as the temperature falls, the mercury is unable to pass the constriction, so it continues to indicate the highest temperature attained.

maximum-wind level The altitude at which the wind speed is greatest.

maximum zonal westerlies The strongest westerly component of the winds that occur in middle and high latitudes.

mean chart A meteorological chart on which the average values for particular features are drawn as isopleths.

mean free path The average distance a molecule travels before colliding with another molecule.

mean temperature The air temperature measured at a particular place over a specified period (e.g., a day, month, or year) and then converted to a mean.

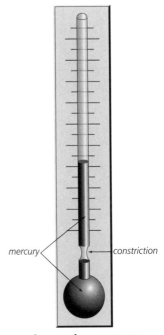

mercury *constriction*

maximum thermometer

mechanical turbulence Turbulent flow produced in moving air that encounters physical obstacles, such as buildings or trees.

medieval warm period A time when the global climate was warmer than it was in the centuries preceding and following it. Temperatures began to rise in about 800 C.E., and as early as 600 C.E. in Greenland, and the peak occurred between 1100 C.E. and 1300 C.E.

mediocris (med) A species of cumulus clouds of moderate vertical extent with fairly small protuberances.

medium-range forecast A weather forecast for a period of five to seven days.

meltemi Northwesterly etesian winds that blow in Turkey, where they often bring relief from the intense summer heat.

melting band A region in certain clouds, especially nimbostratus, where melting snowflakes become coated in a layer of water.

melting level The height at which melting bands occur.

melting point The temperature at which the solid and liquid phases of a substance are in equilibrium at a given pressure. The melting point of pure water at the standard sea-level pressure of 1,013.25 mb (760 mm, 29.92 inches of mercury) is 32°F (0°C, 273.15K). When water melts, 80 cal g^{-1} (334 J g^{-1}) of latent heat is absorbed.

meridional circulation A large-scale movement of air in a north–south or south–north direction, following the lines of longitude (meridians).

meridional flow Movement of air in a north–south or south–north direction on a smaller scale than meridional circulation.

meridional index The component of air movement that is parallel to the lines of longitude (meridians).

meridional wind A wind, or component of a wind, that blows parallel to the lines of longitude (meridians).

mesoclimate The climate of a large area defined by a particular physical characteristic, e.g., a grass-covered plain or mountain range, extending vertically to about 20,000 feet (6,000 m), constitutes a mesoclimate.

mesoclimatology The study of mesoclimates.

mesocyclone A mass of air that spirals upward at the center of a very large cumulonimbus cloud that has developed a supercell.

mesometeorology The study of weather systems that extend horizontally for about 0.6 to 60 miles (1–100 km).

mesopause The boundary separating the mesosphere from the thermosphere. It extends from about 50 miles (80 km) above sea level to a height of about 56 miles (90 km).

mesopeak The temperature maximum, of about 80°F (27°C, 300K), at the stratopause, at a height of about 80 miles (50 km).

mesoscale A scale that extends horizontally from 0.6 to 60 miles (1–100 km).

mesosphere The layer of the atmosphere above the stratopause, extending from about 30 miles (50 km) above the surface to about 50 miles (80 km).

mesotherm A plant that grows in a mesothermal climate.

mesothermal climate A midlatitude climate in which the mean temperature in the coldest month is higher than 26.6°F (–3°C).

Meteor 3 Russian satellite, launched in 1991, that measures the emission and spread of sulfur dioxide from volcanoes.

meteoric water Water that falls from the sky as precipitation.

meteorogram A diagram that shows variable meteorological phenomena plotted against time to illustrate the way weather conditions have changed.

meteorological drought A drought defined as a decrease in precipitation.

meteorological equator The mean latitude of the equatorial trough, which is 5° N. This coincides with the thermal equator.

meteorological minima The lowest values for relevant features of the weather that are prescribed for specified types of flying operations.

meteorology The scientific study of the atmospheric phenomena that produce weather, and especially the application of this study to the forecasting of weather.

Meteosat A European meteorological satellite in geostationary orbit. The first was launched in 1977 by the United States; they are now launched by the European Space Agency. At present there are eight Meteosat satellites in orbit.

methane (CH_4) A greenhouse gas with a global warming potential (GWP) of about 21 that is present in the atmosphere at a concentration of about two parts per million by volume.

methyl chloroform (CH_3CCl_2) A chemical compound once used as a solvent. It is very toxic to humans, has a global warming potential of about 700, and is a source of free chlorine atoms that contribute to the depletion of stratospheric ozone. Its use ended in January 1996.

Meuse Valley incident The first major air pollution disaster. It occurred in 1930 near Liège in the valley of the river Meuse in southern Belgium, when smog collected beneath an inversion. Hundreds of people were ill, more than 60 died, and thousands of cattle were slaughtered.

microbarograph A barograph that records very small changes in atmospheric pressure.

microburst A strong downdraft that occurs below a fairly weak convection cell some distance from the center of a cumulonimbus cloud.

microclimate The climate of a very small area when this can be clearly distinguished from the climate of the surrounding area.

microclimatology The study of microclimates.

micrometeorology The scientific study of the atmospheric conditions inside a microclimate.

microtherm A plant that grows in a microthermal climate.

microthermal climate A midlatitude climate in which the mean temperature in the coldest month is lower than 26.6°F (–3°C); known as a moist subhumid climate in other classifications.

microwave limb sounder (MLS) A spectrometer carried on the Upper Atmosphere Research Satellite that measures concentrations of ozone (O_3), water vapor (H_2O), and chlorine monoxide (ClO) in the stratosphere and O_3 and H_2O in the mesosphere.

microwaves Electromagnetic radiation with a wavelength of 1 mm to 10 cm (0.04–4 in.).

microwave sounding unit (MSU) An instrument carried on the TIROS-N series of NOAA satellites that measures microwave emissions from molecular oxygen in the troposphere, from which atmospheric temperature is calculated with an estimated accuracy of ±0.02°F (±0.01°C). The continuous record of MSU measurements began in January 1979 and shows no statistically significant temperature trend since then.

middle cloud Cloud with a base between 6,500 and 20,000 feet (2,000–6,000 m) altitude. Altocumulus and altostratus are middle clouds.

midlatitude westerlies The prevailing winds between latitudes of about 30° and 60° in both hemispheres.

Mie scattering The scattering of incoming solar radiation when the radiation interacts with particles of a size similar to the wavelength of the radiation. This is predominantly in a forward direction and affects all wavelengths.

migratory An adjective that describes a pressure system embedded in the general westerly airflow of middle latitudes and traveling with it.

Milankovich cycles Variations in the amount of solar radiation the Earth receives due to cyclical changes in the orbit of the Earth about the Sun and in the rotation of the Earth on its axis. There are three such cycles. The first is a variation in the eccentricity of the Earth's orbit, with a period of about 100,000 years. The second is a change from 22.1° to 24.5° and back again in the angle between the Earth's rotational axis and the plane of the ecliptic over about 42,000 years. The third is the precession of the equinoxes over about 25,800 years, due to a wobble in the Earth's rotation. These cycles are believed to trigger the onset and ending of ice ages.

millet rains Heavy rains in East Africa between October and December, when millet is sown.

milli atmospheres centimeter (milli atm cm) A unit that measures volcanic emissions as the thickness of the layer a substance would form if it were the only constituent of the air and subjected to sea-level pressure.

millibar (mb) A unit of atmospheric pressure equal to one-thousandth of a bar. One mb = 0.0145 pounds per square inch = 0.75 mm of mercury (0.03 inches of mercury).

minimum thermometer An alcohol thermometer that records the lowest temperature reached since it was last reset. When the liquid contracts toward the bulb, a small metal strip (the "index") is drawn along the tube by surface tension. When the temperature rises, the liquid expands past the index, leaving the tip farthest from the bulb registering the lowest temperature attained.

mirage An optical phenomenon caused by the refraction of light as it passes from cool to warm or warm to cool air.

mist Liquid precipitation with droplets 0.0002–0.002 inch (0.005–0.05 mm) in diameter. Visibility is more than 1,094 yards (1 km).

mistral A cold, northerly wind over southern Europe bordering the Mediterranean, especially along the lower part of the Rhône River Valley, France.

mixed cloud A cloud containing both water droplets and ice crystals.

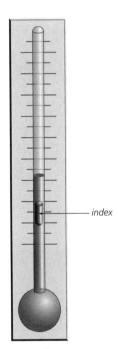

index

minimum thermometer

migratory – mixed cloud

GLOSSARY

mixed nucleus A cloud condensation nucleus consisting of two substances, one hygroscopic and the other possessing the property of wettability.

mixing condensation level The lowest height at which water vapor condenses in a thoroughly mixed layer of air.

mixing depth The distance between the surface of the Earth and the height beyond which vertical air motion by convection ceases.

mixing law When a volume of fluid containing several ingredients is mixed, each ingredient will become distributed evenly throughout the total volume.

mixing length The average distance a particle in an eddy travels perpendicular to the mean path of a turbulent flow of air.

mixing ratio (mass mixing ratio) The ratio of the mass of any gas present in the air to a unit mass of dry air, usually measured as grams of the gas per kilogram of air without the gas and most often used to report humidity as grams H_2O kg^{-1} air. Because it is measured in units of mass, it is not affected by changes of temperature or pressure.

moderate breeze In the Beaufort wind scale, a force 4 wind, which blows at 13–18 mph (21–29 kmh^{-1}).

moderate gale In the Beaufort wind scale, a force 7 wind, which blows at 32–38 mph (51–61 kmh^{-1}).

moist climate A climate in which the annual precipitation exceeds the annual potential evapotranspiration.

moisture factor A measure of the effectiveness of precipitation, calculated by dividing the amount of precipitation in centimeters by the temperature in degrees Celsius for the period under consideration.

moisture inversion A layer of air through which the humidity increases with height.

monsoon A reversal in the direction of the prevailing wind that occurs twice a year over much of the tropics, producing two seasons with distinctly different weather.

monsoon fog A fog that occurs along some coasts during the summer (wet) monsoon.

monsoon low An area of low surface pressure over a continent during the summer and over the adjacent ocean during the winter; the pressure distribution associated with monsoon conditions.

monsoon trough A trough that forms during the summer (wet) monsoon at about latitude 25° N and extending just to the south of the Himalayas.

month degrees A measure of the conditions for plant growth used in some climate classifications and calculated by subtracting 43°F (6°C) from the mean temperature for each month and adding together the remainders (the number of degrees by which the temperature is above or below 43°F), 43°F being the minimum temperature for growth of most plant species.

Montreal Protocol on Substances that Deplete the Ozone Layer An international agreement reached in 1987 that aims to reduce and eventually eliminate the release into the atmosphere of all man-made substances that deplete stratospheric ozone.

moon dog A bright patch occasionally seen to one side of the Moon, usually slightly below it, caused by the refraction of light by ice crystals that are falling very slowly.

moon pillar A column of light occasionally seen above or below the Moon, when the Moon is low in the sky.

mother cloud A cloud from which other clouds have been produced and that is seen at the same time as the clouds to which it gave rise.

Mount Agung (Gunung Agung) A volcano on the island of Bali, Indonesia, that erupted violently in 1963, ejecting large amounts of particulate material, some of which entered the lower stratosphere. The absorption of solar energy by the stratospheric aerosols warmed the lower stratosphere by 11–12°F (6–7°C) and surface temperatures fell by about 1°F (0.5°C).

mountain breeze A katabatic wind that blows at night in some mountain regions.

mountain climate A climate differing from the climate typical of the latitude because of the high elevation. Generally, mean temperatures are lower, conditions are windier, and precipitation is greater than at lower levels but decreases above the permanent snow line, because the air has lost most of its moisture.

mountain breeze

mountain-gap wind (canyon wind, gorge wind, jet-effect wind) A local wind that occurs where the prevailing wind is funneled between two mountains and accelerated.

mountain meteorology The study of the effects mountains have on the atmosphere and the weather conditions that result.

Mount Aso-san A volcano on the Japanese island of Kyushu that erupted violently in 1783, injecting a large amount of material into the stratosphere. The eruption was followed by unusually cold weather from 1784 to 1786.

Mount Katmai A volcano in Alaska that erupted violently in 1912. During the months that followed there was a 20% drop in solar radiation and the weather was unusually cool, although temperatures had been somewhat lower than normal before the eruption.

Mount Pinatubo A volcano on the island of Luzon, Philippines, that erupted in 1991, ejecting dust and sulfate aerosol to a height of 25 miles (40 km). The resulting absorption of solar radiation caused a cooling of about 0.7–1.25°F (0.4–0.7°C) that lasted through 1992 and 1993.

Mount Spurr A volcano in Alaska that erupted in 1992. There is no record of any climatic effect.

Mount Tambora A volcano in the Dutch East Indies (now Indonesia) that erupted in 1815, ejecting dust and sulfuric acid aerosols that spread to form a veil over much of the Northern Hemisphere. Temperatures fell by up to 2°F (1°C) in many areas and wind patterns were distorted. The following year, 1816, came to be known as "the year with no summer."

mud rain Rain containing fine soil particles that discolor the water and leave a mud-like deposit on surfaces.

multichannel sea surface temperature (MCSST) A procedure in which sea-surface temperatures are calculated from data received from an advanced, very high resolution radiometer.

multispectral scanner An instrument carried on Landsat satellites that obtains images of the Earth's surface used to monitor surface changes, for example, in vegetation, coastlines, ice sheets, glaciers, and volcanoes.

mutatus A cloud development in which the cloud shape changes fairly rapidly because of new cloud masses that are growing from it.

nacreous cloud (mother-of-pearl cloud) Bright, white clouds, usually tinged with pink, occasionally seen in high latitudes shortly before dawn or after sunset.

nadir The point on the surface lying directly beneath an observational satellite.

n'aschi A northeasterly wind that blows in winter along the coasts of Iran and Pakistan.

natural seasons Five periods, each lasting for at least 25 days, that are characterized by weather of a distinct type. Spring–early summer lasts from early April until the middle of June. High summer lasts from the middle of June until early September. Autumn lasts from the middle of September until the middle of November. Early winter lasts from the third week in November until the middle of January. Late winter–early spring last from the third week in January until the end of March.

navigable semicircle The side of a tropical cyclone closest to the equator, where the winds are lightest and tend to push ships out of the path of the storm.

near-infrared radiation Infrared radiation that has the shortest wavelength in the infrared waveband, of about 1–3 μm.

near-infrared mapping spectrometer An instrument carried on some satellites that takes readings at near-infrared wavelengths, from which the chemical composition, structure, and temperature of planetary and satellite atmospheres can be calculated, as well as details of the surface mineral and geochemical composition.

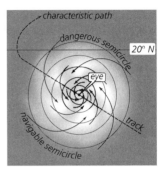

navigable semicircle

near UV Ultraviolet radiation at wavelengths of 400–300 nm.

nebulosus (neb) A species of clouds that form a layer or veil with no clearly distinguishable features.

neon (Ne) One of the noble gases that is colorless and is present in the atmosphere, accounting for 0.00182% by volume. The atmosphere of the other planets and satellites have much less neon. It vaporizes at −410.89°F (−246.05°C) and freezes at −416.61°F (−248.67°C).

nephanalysis The study of cloud types and formations in satellite images to obtain information about the weather systems that produce them.

nephcurve A line drawn to mark the boundary between clouds and clear sky, areas of precipitation, clouds of different types, or clouds at different heights.

nephelometer A laser instrument used in remote sensing to measure the scattering of light by atmospheric particles.

nephology The scientific study of clouds.

nephoscope An instrument used to measure the direction of movement of a cloud and its angular velocity around a point on the surface directly beneath it.

neutral atmosphere An atmosphere in which the environmental lapse rate (ELR) and dry adiabatic lapse rate (DALR) are equal.

neutral stability The condition in which the atmosphere is stratified and the potential temperature remains constant with height.

neutrosphere The atmosphere from the surface to the base of the ionosphere, at about 43–56 miles (70–90 km). Within this shell most atmospheric constituents are electrically neutral.

nevados A cold, katabatic wind that blows down the mountainsides and into the high valleys of Ecuador.

Next Generation Weather Radar (NEXRAD) A network of 175 Doppler radar installations operated by the National Weather Service at weather stations, airports, and military airfields throughout the United States.

nightglow Radiation emitted at night in the mesosphere, mainly 50–60 miles (80–100 km) above the surface, due to chemical reactions among atmospheric gases. It is too faint to be visible from the ground, but satellite instruments can detect it.

nimbostratus (Ns) A genus of low, gray, fairly uniform clouds that often deliver steady, continuous rain or snow.

nimbus Latin for "cloud." Attached to the cloud genus stratus, to give nimbostratus, and the genus cumulus to give cumulonimbus, in both cases associating the name with the idea of precipitation.

Nimbus satellites A series of U.S. weather satellites that carry equipment for Automatic Picture Transmission, a television camera system for mapping clouds, and an infrared radiometer that allows pictures to be taken at night.

nitrogen (N) A colorless gas that accounts for 78.08% of the atmosphere by volume. It freezes at –345.75°F (–209.86°C) and boils at –320.44°F (–195.80°C). Nitrogen is an essential component of all proteins and nucleic acids.

nitrogen cycle The sequence of reactions by which nitrogen moves between the atmosphere, soil, living organisms, and returns to the air.

nitrogen oxides (NO_x) Any of the seven oxides of nitrogen, but specifically nitric oxide or nitrogen oxide (NO) and nitrogen dioxide (NO_2), both released mainly by burning fossil fuels and plant material, especially vehicle exhausts and the burning of waste following forest clearance in the Tropics. These gases are catalysts for the formation of ozone in the troposphere, playing a critical role in the reactions that lead to the formation of photochemical smog, but can cause the destruction of ozone in the stratosphere.

nitrous oxide (N_2O) A greenhouse gas with a global warming potential of 310 that is released by bacteria from soils in wet tropical forests and dry savannah grasslands and from the oceans. Certain industrial processes and automobiles fitted with three-way catalytic converters also emit it.

noble gases (inert gases) The chemical elements helium (He), neon (Ne), argon (Ar), krypton (Kr), xenon (Xe), and radon (Rn), which together comprise group 0 (or group VIII) of the periodic table.

noctilucent cloud Cloud at a height of about 50 miles (80 km), seen occasionally on summer nights in high latitudes shining with light reflected from the Sun.

nomogram A diagram showing the relationship between two or more values or scales of measurement, or two types of measurement, such as the change of atmospheric pressure with altitude.

non-frontal depression An area of low atmospheric pressure that is not associated with a frontal system.

non-selective scattering The scattering of all wavelengths of light equally, caused by particles bigger than the wavelength of the radiation. This produces a white sky.

Nordenskjöld line A line drawn to link places where the mean temperature in the warmest month is high enough for trees to grow and the mean temperature in the coldest is not low enough to kill trees. This marks the boundary between boreal forest and tundra.

nor'easter (northeast storm) A storm with northeasterly winds of up to hurricane force (75 mph, 121 kmh^{-1}), in eastern North America from Virginia to the Canadian Maritime Provinces.

normal The mean value of any meteorological value, calculated from measurements made at a particular place over a long period.

normal chart (normal map) A chart on which the distribution of normal weather features is plotted.

normalize To produce a dimensionless ratio between two quantities by dividing one quantity by a more fundamental quantity in the same dimensions.

Normalized Difference Vegetation Index (NDVI) An index that measures the amount of actively photosynthesizing plant biomass in a landscape; shown on maps compiled from satellite data.

norte (papagayo) A cold, northerly wind that blows in winter down the eastern coast of Mexico.

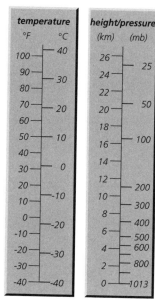

nomogram

North American high A weak area of high surface pressure that covers most of North America in winter.

North Atlantic oscillation (NAO) A periodic change in the distribution of sea-level atmospheric pressure between Iceland and the Azores. When pressure is lower than average over Iceland and higher than average over the Azores, the index is high. When the Iceland low and Azores high are both weaker than average, the index is low.

norther A strong, cold winter wind from the north that blows across the southeastern United States, sometimes extending across the Gulf of Mexico and Central America into the Pacific.

northern circuit The path depressions usually follow in summer as they cross the United States from west to east.

Northern Rocky Mountains wind scale (NRM wind scale) A scale of wind strength adapted by the U.S. Forest Service from the Beaufort wind scale for use in the forests of the northern Rocky Mountains.

north föhn A föhn wind generated by a movement of air from north to south across the Alps.

north temperate zone That part of the Earth lying between the tropic of Cancer and the Arctic Circle.

nor'wester (northwester) Any wind or weather system that arrives from the northwest. In South Island, New Zealand, a hot, dry wind from the mountains. In northern India, a storm caused by convection, especially in the Ganges Delta prior to the break of the summer monsoon. In South Africa, a depression associated with an active front.

nowcasting The issuing of local weather forecasts for up to two hours ahead.

numerical forecasting A method of weather forecasting that begins with detailed measurements of the state of the atmosphere at many different places at regular intervals. These reveal the way conditions are changing. The changes are interpreted mathematically, by applying equations derived from known physical laws, and projected forward to produce the forecast.

Nusselt number A dimensionless number used in calculations of heat transfer in fluids. It is the ratio of heat transferred to the amount that would be transferred by pure conduction under ideal conditions.

oasis effect The cooling effect that produces a difference in local weather found where an area of moist ground (an "oasis") is surrounded by dry ground ("desert"). It is due to cooling caused by evaporation.

obscuration The situation when the sky is completely hidden by a weather feature at ground level, such as fog.

obscuring phenomenon Any atmospheric feature, other than cloud, that obscures a portion of the sky as seen from a weather station.

obstruction to vision An atmospheric feature that reduces horizontal visibility at ground level.

occlusion The stage in the life cycle of a frontal system at which the advancing cold air has started to lift the warm air clear of the surface. This is shown on weather maps as a front with alternating triangles and semicircles.

occultation The passing of one celestial object in front of another, so that one of the objects is partly or completely hidden to an observer.

occult deposition The depositing of acid onto surfaces when they come into direct contact with mist or cloud droplets containing dissolved acid.

oceanicity The extent to which the climate in a particular place resembles the most extreme type of maritime climate.

ocean weather station A weather station carried on a ship anchored in a specified location at sea.

offshore wind A wind that blows from the land in the direction of the sea.

okta (octa) A unit used to report the extent of cloud cover in eighths of the total sky. Symbols for cloud amount are in oktas, but can be interpreted as: 1/8 = 1/10 or less; 2/8 = 2/10–3/10; 3/8 = 4/10; 4/8 = 5/10; 5/8 = 6/10; 6/8 = 7/10–8/10; 7/8 = 9/10 or overcast but with gaps in the cloud cover.

Older Dryas A cold period that occurred in northern Europe from about 12,200 years ago to about 11,800 years ago, soon after the ice sheets had retreated at the end of the Devensian glacial.

onshore wind A wind that blows from the sea in the direction of the land.

opacus A variety of clouds that form a layer, sheet, or patch that hides the Sun or Moon completely.

opaque sky cover The proportion of the sky covered by cloud that completely hides anything that might be above it.

Operational Linescan System (OLS) A satellite imaging system that detects lightning and waste gas flares at oil wells. It scans the whole Earth once every day at visible and infrared wavelengths.

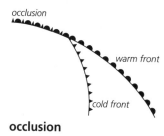

occlusion

occlusion

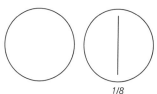

1/8

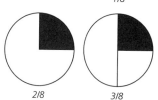

2/8   3/8

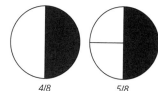

4/8   5/8

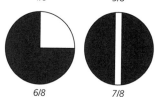

6/8   7/8

overcast   sky obscured

okta

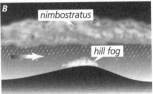

orographic cloud

optical air mass (airpath) The length of the path through the atmosphere that light from the Sun or any other celestial body travels.

optical depth (optical thickness) A measure of the extent to which a cloud or atmospheric layer prevents the vertical passage of solar radiation.

orbital forcing Changes in climate due to variations in the Earth's orbit around the Sun.

orographic cloud Cloud that forms above high ground as a result of orographic lifting. Small cumulus clouds form if the air is fairly buoyant. If the air is moist, it forms a thicker layer of nimbostratus with a cap of stratus sitting as hill fog on the highest ground. Lenticular cloud is also a form of orographic cloud.

orographic lifting The forced raising of air as it crosses high ground.

orographic occlusion An occlusion that forms when an advancing warm front reaches a mountain range. The front slows, causing warm air to accumulate and allowing the cold front to undercut it more rapidly than it would have done over level ground.

orographic rain Rain that falls on the windward slope of a hill or mountain as a direct consequence of orographic lifting.

oscillatory wave A wave that causes air or water to move about a point, but without advancing in the direction the wave is moving.

ouari A southerly wind that blows across Somalia, Africa.

outburst A sudden, very heavy fall of precipitation caused by the strong downcurrents in a cumuliform cloud.

outflow boundary The location in a thunderstorm where air that has been cooled by the evaporation of moisture meets warm, moist air.

overcast The condition when cloud covers at least 7 oktas (9/10) of the sky.

overrunning The situation in which warm air rides up a frontal surface, moving over the cold air beneath the front.

overseeding Cloud seeding in which an excessive amount of nucleating material is delivered into the cloud. A much larger number of ice crystals form, but they are too small to fall. Consequently, overseeding inhibits precipitation.

oxygen (O) A colorless, odorless gas comprising 20.946% of the atmosphere by volume, with a concentration of 209,460 parts per million. Its atomic number is 8, relative atomic mass 15.9994, melting point

–353.92°F (–214.4°C), and boiling point –297.4°F (–183°C). In its commonest form oxygen atoms bond in pairs (O_2), but atoms can also bond in threes to form ozone (O_3). Oxygen is extremely reactive and many oxidation reactions release energy. Aerobic respiration releases energy from oxidation reactions and combustion releases heat through chain reactions in which carbon and hydrogen are oxidized. Oxygen is slightly soluble in water, and aquatic aerobic organisms, such as fish, use dissolved oxygen for respiration.

oxygen isotope ratios The ratio of the two common isotopes of oxygen, ^{16}O and ^{18}O, in substances recovered from sediments or ice cores that can be dated. The ratio provides information about past climates. An increase in the proportion of ^{18}O in calcium carbonate indicates climatic cooling and an increase of ^{18}O in ice indicates warming.

ozone A pale blue gas and powerful oxidizing agent that is a form of oxygen in which the molecule comprises three atoms (O_3) rather than the two of ordinary oxygen (O_2).

ozone depletion potential A measure of the extent to which a particular substance is likely to remove ozone from the ozone layer, compared with the extent to which CFC-11 (Freon-11) does so. CFC-11 is given a value of 1 and other compounds are evaluated on this scale.

ozone hole The area over Antarctica in which the concentration of stratospheric ozone decreases during late winter and early spring.

ozone layer A region of the stratosphere between 66,000 and 98,000 feet (20 km and 30 km) above the ground where the concentration of ozone (O_3) is usually higher than it is elsewhere, commonly reaching 10^{18}–10^{19} molecules per cubic meter, or 220–460 Dobson units.

Pacific air Maritime air that has been modified as it crossed the Rocky Mountains.

Pacific decadal oscillation (PDO) A change that occurs over a period of several decades in the ocean–atmosphere system in the Pacific Basin, affecting the temperature of the lower atmosphere.

Pacific highs Two anticyclones covering a large part of the subtropical North and South Pacific Oceans.

paleoclimatology The scientific study of the climates that existed in the distant past.

Palmer drought severity index (PDSI) An index for classifying droughts that measures the extent to which the water supply departs from what is normal for a particular place. It is widely used in the United States for monitoring droughts.

pampero A violent squall that blows from the southwest across the pampas of Argentina and Uruguay, north of the River Plate.

panas oetara A strong, warm, and dry northerly wind that blows in Indonesia in February.

pan coefficient The ratio of the amount of water that evaporates from a unit area of the exposed surface of a large body of water to the amount that evaporates from an evaporation pan of similar area over a similar period.

pannus An accessory cloud comprising ragged patches of cloud attached to or beneath another cloud.

parcel method A way of testing the stability of air based on calculating the consequences of displacing particular parcels of air.

parcel of air (air parcel) A volume of air that can be considered separately from the air surrounding it and from which it is assumed to be physically isolated.

partial obscuration The condition of the sky when up to 7 oktas (9/10) is hidden by an obscuring phenomenon on the surface.

partial pressure In a mixture of gases, the share of the total pressure that can be attributed to one of the constituent gases.

particulate matter Fine particles suspended in the atmosphere.

partly cloudy The condition of the sky when the average cloud cover over a period of 24 hours has been 1–4 oktas (10–50%).

passive instrument An instrument that measures radiation falling on it (so it is passive), rather than sending out a signal that returns to it.

past weather The weather that has prevailed over the period, usually of six hours, since a weather station last submitted a report.

path length The distance incoming solar radiation must travel between the top of the atmosphere and the land or sea surface. The path length has a value of 1 when the Sun is directly overhead, or 90° above the horizon, and it increases as the angle of the Sun decreases. It can therefore be calculated as 1/cosA, where A is equal to 90° minus the angle of the Sun.

peak gust The highest wind speed (sustained or gust) recorded at a weather station during a period of observation, commonly of 24 hours.

pearl-necklace lightning (beaded lightning, chain lightning) A rare type of lightning that appears as a chain of lights, or a string of pearls.

pea souper Yellowish smog containing more smoke than fog in which visibility is less than 30 feet (10 m).

penetrative convection The type of convection that occurs when the ground or water surface is warmer than the air immediately above it. The air is warmed from below, producing conditional instability leading to the formation of cumuliform clouds and chaotic turbulence.

Penman formula A mathematical formula for measuring the rate of evaporation from an open surface, expressing evaporation losses in terms of the duration of sunshine, mean air temperature, mean humidity, and mean wind speed.

pennant A triangular symbol, resembling a pennant flag, used on synoptic charts to indicate wind speeds greater than 48 knots (55 mph, 89 kmh^{-1}).

pentad A period of five days often used in preference for the week, because there is an exact number of pentads (73) in a 365-day year.

penumbra The less deeply shaded area that lies near the edge of a shadow.

periglacial climate The climate that prevails near the edge of an ice sheet.

perihelion The point in the eccentric solar orbit of a planet or other body when it is closest to the Sun. At present, the Earth is at perihelion on about January 3. The dates of aphelion and perihelion change over a cycle of about 21,000 years.

period The amount of time that elapses between two events, and in particular between two repetitions of the same event. The period of a wave is the time that elapses between two wave crests (or troughs) passing a fixed point.

perlucidus A variety of clouds that form an extensive layer, sheet, or patch with open spaces.

permafrost (pergelisol) Ground that remains frozen throughout the year. In order for permafrost to form the temperature must remain below freezing for at least two consecutive winters and the whole of the intervening summer.

permanent drought The type of drought characterizing deserts. There are no permanent streams or rivers; precipitation is rare, although often heavy when it occurs; and crops can be grown only on irrigated land.

peroxyacetyl nitrate (PAN) A chemical compound ($CH_3CO.O_2NO_2$) that forms by a complicated series of reactions involving the oxidation of

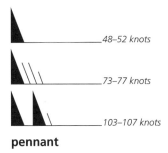

48–52 knots

73–77 knots

103–107 knots

pennant

hydrocarbons, especially in vehicle exhausts. In warm air PAN decomposes to release nitrogen dioxide (NO_2) and the highly reactive peroxyacetyl radical $CH_3CO.O_2$. This process contributes to the atmospheric content of nitrogen oxides (NO_x), implicated in the formation of photochemical smog and acid rain.

persistence The length of time a particular weather feature remains unchanged.

persistence forecast A weather forecast that predicts a continuation of present conditions for several hours ahead.

phase (1) A part of a system that is of the same composition throughout and distinct from all other parts of the system. When water changes between gas, liquid, and solid, it is said to change phase. (2) The stage a regularly repeating motion has reached.

phenology The study of periodic events in the lives of plants and animals that are related to the climate.

photochemical smog A form of air pollution that occurs when ultraviolet radiation in strong sunlight acts upon hydrocarbon compounds emitted by vehicle exhausts.

photochemistry The branch of chemistry concerned with the chemical effects of electromagnetic radiation.

photodissociation The splitting of a molecule into smaller molecules or single atoms using light as a source of energy.

photoionization The ionization of an atom that occurs when it absorbs a photon of electromagnetic radiation with a wavelength of less than about 0.1 μm. Photoionization occurs in the ionosphere.

photolytic cycle A naturally occurring sequence of reactions in which ultraviolet radiation supplies the energy for the photodissociation of nitrogen dioxide (NO_2). The NO_2 then reforms. The reactions are:

$$NO_2 + UV \rightarrow NO + O \qquad (1)$$
$$O + O_2 \rightarrow O_3 \qquad (2)$$
$$O_3 + NO \rightarrow NO_2 + O_2 \qquad (3)$$

photoperiod The number of hours of daylight that occur during a 24-hour period.

photopolarimeter-radiometer (PPR) An instrument used in remote sensing that measures the intensity and polarization of sunlight in the visible part of the spectrum, from which the temperature and cloud

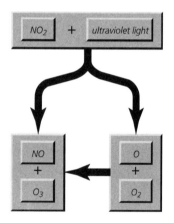

NO_2 = nitrogen dioxide
NO = nitrogen oxide
O = oxygen atom
O_2 = oxygen molecule
O_3 = ozone

photolytic cycle

formation in the atmosphere of a planet or satellite can be determined, as well as some surface detail.

photosphere The visible surface of the Sun or any other star.

physical climatology The branch of climatology dealing with exchanges of mass and energy in the troposphere, i.e., with the physical processes that produce and regulate climate.

physical meteorology The branch of meteorology concerned with the physical processes involved in producing the day-to-day weather.

physisorption The physical bonding of molecules by van der Waals' forces to the surface of a solid object or, less commonly, of a liquid.

phytoclimatology The study of the climatic conditions in the air between and adjacent to growing plants and on plant surfaces.

Picasso-CENA (Pathfinder Instruments for Cloud and Aerosol Spaceborne Observations—Climatologie Etendue des Nuages et des Aerosols) A joint U.S.–French satellite to be launched in 2003 that will measure aerosols.

piezoresistance barometer A barometer based on elements that measure electrical resistance. Piezoresistance barometers are accurate, small, light, robust, and inexpensive, which makes them convenient to carry and use on vehicles and radiosondes.

pileus An accessory cloud that forms a smooth, thin covering above or attached to the top of a cumuliform cloud, like a cap or hood.

pilot balloon (pibal) A weather balloon that ascends at a predetermined rate. As it rises, the balloon is tracked, and at intervals its altitude is calculated from its known rate of ascent and its azimuth angle. The wind velocity is then calculated for each height. If the sky is obscured, the height of the cloud base can be measured.

pilot report A description of the current weather conditions that is radioed to air traffic control or to a meteorological center by the pilot or other crew member of an aircraft.

pitot tube (pitot–static tube) A device used to sample air in order to measure wind speed or airspeed and atmospheric pressure. Pitot tubes are used at weather stations, but their most widespread use is on aircraft. Every aircraft carries a pitot tube connected to its altimeter, airspeed indicator, and vertical speed indicator.

plage A bright area on the chromosphere of the Sun.

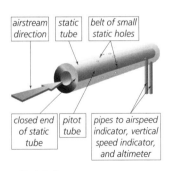

pitot tube

Planck's law A law stating that the intensity of radiation emitted by a body at a given wavelength is determined by the temperature of the emitting body. This can be written as $E_\lambda = c_1/[\lambda^5 (\exp(c_2/\lambda T) - 1)]$, where E_λ is the amount of energy (in watts per square meter per μm of wavelength), λ is the wavelength (in μm), T is the temperature in kelvins, c_1 is the first radiation constant (= 3.74×10^{16} W m^{-2}), and c_2 is the second radiation constant (= 1.44×10^{-2} m K).

plane of the ecliptic An imaginary disk, the circumference of which is defined by the Earth's orbital path about the Sun.

planetary boundary layer (atmospheric boundary layer, surface boundary layer) The lowest part of the atmosphere, where the movement of air is strongly influenced by the land or sea surface. Its depth varies, but is usually less than about 1,700 feet (519 m).

planetary vorticity (*f*) The vorticity about a vertical axis that a mass of fluid moving on the surface of the Earth possesses by virtue of the Earth's rotation. Its magnitude is equal to that of the Coriolis effect.

planetary wind Any wind caused wholly by the interaction of solar radiation and the rotation of the Earth.

plane wave A wave front that is not curved owing to its distance from the source.

plume rise The height to which a chimney plume rises.

pluvial A prolonged period of increased precipitation affecting a large region.

point discharge An upward flow of positive ions from tall objects that is induced by the negative charge at the base of a thunderstorm cloud.

point rainfall The amount of rain falling into a particular rain gauge over a specified period, or the amount estimated to have fallen during that period at a particular place.

Poisson's equation An equation from which it is possible to calculate the temperature at any height of a parcel of air on a dry adiabat. The equation is $(T \div \Phi) \times c_p \div R = p \div p_o$, where T is the actual temperature, Φ is the potential temperature, R is the specific gas constant of air, c_p is the specific heat at constant pressure, p is the pressure at the position of the air parcel, and p_o is the surface pressure.

polacke A cold, dry katabatic wind that blows in winter over northern Bohemia, in the Czech Republic.

polar air Cold air that originates in the high-pressure regions of Siberia, northern Canada, and the Southern Ocean.

polar-air depression A non-frontal depression that occurs in the Northern Hemisphere when unstable arctic or polar maritime air moves southward along the eastern side of a large north–south ridge.

polar automatic weather station An automatic weather station designed to operate in extremely cold climates.

polar cell That part of the atmospheric circulation in which air subsides over high latitudes, flows away from the pole at low level, rises at the polar front where it meets the Ferrel cell, then flows back toward the poles.

polar climate A high-latitude climate in which the mean monthly temperature remains below freezing throughout the year.

polar desert An area inside the Arctic or Antarctic Circle where the annual precipitation is very low.

polar easterlies The prevailing winds in the polar cell, caused by a Coriolis effect acting on air flowing away from the pole.

polar-easterlies index A scale of values, expressed as the east-to-west component of the geostrophic wind, that allows the strength of the polar easterlies to be compared.

polar front The front in middle latitudes that marks the boundary between polar and tropical air.

polar front jet stream The jet stream associated with the polar front.

polar high The persistent region of high surface atmospheric pressure covering the Arctic Basin and Antarctica.

polar hurricane A storm with hurricane-force winds generated by an intense polar low.

polar low A small, intense cyclone that forms during winter in the cold air on the side of the polar front nearest the pole.

polar mesospheric cloud The preferred name for noctilucent cloud, indicating that it is seen only in polar latitudes and occurs in the upper mesosphere.

polar night vortex A vortex that forms in winter in the polar stratosphere with a strongly geostrophic wind circulating around it, within which polar stratospheric clouds form.

polar orbit A satellite orbit that passes close to the North and South Poles at an altitude of about 534 miles (860 km).

polar outbreak An extension of polar air into lower latitudes.

polar stratospheric clouds (PSCs) Clouds that form in winter over Antarctica and, less commonly, over the Arctic at a height of 9–15.5 miles (15–25 km). Usually too thin to be visible, when the Sun is about 5° below the horizon they can sometimes be seen as nacreous clouds.

polar trough A trough in the upper troposphere that extends toward the equator far enough to reach the Tropics.

polar vortex The large-scale circulation dominating the middle and upper troposphere in high latitudes, centered in polar regions of both hemispheres.

Pole of Inaccessibility The point in the Arctic Ocean, between the North Pole and Wrangel Island, that is farthest from land. It is sometimes taken to be the center of the Arctic.

pollution Any direct or indirect alteration of the properties of any part of the environment in such a way as to present an immediate or potential risk to the health, safety, or well-being of any living species. The alteration may be chemical, thermal, biological, or radioactive.

Pollution Standards Index (PSI) An internationally agreed upon measure of air quality that compares the national air-quality standard with the amount of the principal pollutant. If the amount is less than the national standard, the PSI value is less than 100 and air quality is moderate or good. If the amount is greater than the national standard, the PSI value is more than 100 and the air quality is poor. A PSI of 200 is considered "very unhealthy," above 300 is "hazardous," and above 400 is "very hazardous."

polyn'ya An area of open water surrounded by sea ice.

ponding The accumulation of cold air in a frost hollow.

ponente A westerly wind along the Mediterranean coast of France and in Corsica.

poniente A westerly wind through the Straits of Gibraltar.

poriaz (poriza) A strong, northeasterly wind across the Black Sea in the region of the Bosporus.

port meteorological officer (PMO) An official belonging to a national meteorological service who supervises the voluntary observing ships (VOS) scheme.

positive axis A line drawn through the point of maximum curvature in the streamline of an easterly wave.

post-glacial climatic revertence A period when the climate became cooler and wetter than it had been previously. The change began abruptly about 2,500 years ago and the cool, wet conditions continue to the present day.

potential energy The energy stored in a body by virtue of its position or state.

potential evapotranspiration The amount of water that would leave the ground surface by evaporation and transpiration if the supply of water were unlimited.

potential temperature (Φ) The temperature a volume of a fluid would have if the pressure under which it is held were adjusted to sea-level pressure and its temperature were to change adiabatically.

potential temperature gradient The difference between the adiabatic lapse rate and the environmental lapse rate (ELR).

power-law profile A mathematical expression describing the variation of the wind with height.

Poza Rica incident An industrial accident in 1950 at Poza Rica, Mexico. Equipment failure at the sulfur-recovery unit of an oil refinery caused the release of hydrogen sulfide (H_2S), which was trapped beneath an inversion. About 320 people became ill and 22 people died.

praecipitatio A supplementary cloud feature consisting of precipitation falling from the cloud and appearing to reach the ground.

precession of the equinoxes A change in the dates at which the Earth reaches aphelion and perihelion and therefore in its position in its orbit at the equinoxes and solstices.

precipitable water vapor The total amount of water vapor present in a column of air above a point on the surface of the Earth, or in a column of air within an atmospheric layer defined by the atmospheric pressure at its base and top.

precipitation Water that falls from the sky to the surface in either liquid or solid form, including fog, dew, and frost.

precipitation area The area on a synoptic chart over which precipitation is falling, often shown by shading on TV weather maps.

precipitation ceiling The vertical visibility measured looking upward into precipitation.

precipitation cell An area of fairly continuous precipitation indicated by radar.

precipitation current A downward flow of electric charge caused by falling precipitation.

precipitation echo A radar image caused by the reflection from precipitation.

precipitation-efficiency index A value indicating the amount of water available for plant growth. It is given by $115(r/t - 10)^{10/9}$, where r is the mean monthly rainfall in inches and t is the mean monthly temperature in degrees Fahrenheit. This calculation is made for each month, and the sum of the indexes for 12 months gives the precipitation-efficiency index.

precipitation-generating element A small region inside a cloud where ice crystals are growing more rapidly than elsewhere at the expense of supercooled water droplets.

precipitation intensity The amount of precipitation falling to the surface within a specified period, measured in inches or millimeters per hour or day.

precipitation inversion (rainfall inversion) An inversion that inhibits precipitation by restricting the vertical movement of air rising by convection, thus reducing the amount of condensation and consequent precipitation.

precipitation physics The branch of physical meteorology concerned with the processes involved in the formation of cloud droplets, ice crystals, and the resulting precipitation.

precipitation station A weather station where only the amount and type of precipitation is measured and recorded.

prefrontal surge The descent of dry air over an occlusion and ahead of the upper-level front.

present weather The current weather conditions included in a report from a weather station, represented by a series of two-digit numbers, from 00 to 99.

pressure altitude The height above sea level at which the air pressure in a standard atmosphere would be the same as the pressure measured at the surface in a particular place. The use of pressure altitudes allows atmospheric pressure to be expressed in terms of altitude.

pressure altitude variation The difference between the true altitude and pressure altitude.

pressure anemometer An anemometer that uses a pitot tube to measure the pressure exerted by the wind; this is converted to wind speed.

pressure center The center of an area of low or high pressure as these appear on a synoptic chart or weather map.

pressure-change chart (pressure-tendency chart) A chart showing the change in air pressure over a specified period across a surface at a constant height.

pressure-fall center (isallobaric low, katabaric center) The place where the air pressure has fallen further than anywhere else over a specified period.

pressure gradient (isobaric slope) The rate at which atmospheric pressure changes over a horizontal distance. A line drawn at right angles to the isobars shows the direction of the gradient, and the distance between isobars indicates the steepness of the gradient.

pressure-gradient force (PGF) The force produced by the pressure gradient, acting at right angles to the pressure gradient with a magnitude proportional to the steepness of the gradient.

pressure pattern The distribution of air pressure as shown by the isobars on a synoptic chart.

pressure-plate anemometer An anemometer consisting of a flat plate that swings freely at the end of a horizontal arm. Air pressure makes the plate swing inward and the distance it swings is converted into wind speed.

pressure-rise center (anabaric center, isallobaric high) The place where the air pressure has risen further than anywhere else over a specified period.

pressure system An atmospheric feature characterized by air pressure.

prester (1) A very hot whirlwind accompanied by lightning in the Mediterranean region and especially in Greece. (2) A waterspout accompanied by lightning.

prevailing visibility The greatest horizontal visibility extending over at least one-half of the horizon around an observation point.

prevailing westerlies (westerlies) The westerly winds that predominate in middle latitudes. They are strongest at about latitudes 35–40° N and S.

prevailing wind The direction from which the wind at a particular place blows more frequently than from any other.

primary circulation That part of the general circulation of the atmosphere comprising large-scale, persistent features.

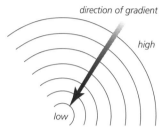

pressure gradient

primary cyclone (primary low) An area of low atmospheric pressure (cyclone) that has one or more smaller (secondary) cyclones within its circulation.

primary front The first front to form in a frontal system.

primary pollutant A substance released into the environment where it causes immediate pollution.

probability forecast A weather forecast that states the expected likelihood of a particular type of weather, usually precipitation.

probable maximum precipitation An estimate of the greatest amount of precipitation that could conceivably fall on a given drainage area over a given period, calculated from records of the worst storms known in the area.

prognostic chart (forecast chart) A synoptic chart showing weather features, sometimes including the position of fronts, as they are expected to appear at some specified time in the future.

progressive wave A wave or group of waves that move in relation to the surface of the Earth.

proxy data Data that do not refer to the climate directly, but can be interpreted to yield information about climate.

pseudoadiabat A line drawn on a thermodynamic diagram to show the lapse rate in air that is rising past the lifting condensation level.

pseudofront A boundary, resembling a small front, between air cooled by rain falling through a large cumulonimbus cloud and the warmer air adjacent to it.

psychrometer A hygrometer that uses two thermometers. The dry-bulb thermometer measures air temperature; the wet-bulb thermometer indirectly measures the rate of evaporation. The bulb of this thermometer is wrapped in a wick, partly immersed in a reservoir of distilled water.

psychrometric equation An equation used to calculate the specific humidity of air from the wet-bulb and dry-bulb temperatures: $q(T) = q_s(T_w) - \lambda(T - T_w)$, where $q(T)$ is the specific humidity of the air, $q_s(T_w)$ is the saturated specific humidity of the air, T is the dry-bulb temperature, T_w is the wet-bulb temperature, and λ is the psychrometric constant (= C_p/L, where C_p is the specific heat of air and L is the latent heat of vaporization of water). The value of the psychrometric constant varies according to the temperature and atmospheric pressure. At 68°F (20°C) and a pressure of 14 lb. in.$^{-2}$

dry-bulb thermometer   wet-bulb thermometer

wick

reservoir

psychrometer

(100 kPa), it is 0.0009 oz. ft.$^{-3}$ °F^{-1} (0.489 g m^{-3} K^{-1}). The psychrometric equation can be converted to calculate the vapor pressure or vapor density of the air.

psychrometry The measurement of dry-bulb and wet-bulb temperatures using a psychrometer and calculating the specific humidity, vapor pressure, or vapor density of the air from those values.

puelche (fog wind) A föhn wind on the western side of the Andes.

puff of wind A breeze strong enough to produce a patch of ripples on the surface of still water.

pulse radar Radar transmitted as short, intense bursts of radiation with a fairly long interval between bursts.

purga A strong type of blizzard in the tundra of northeastern Siberia.

pyranometer (solarimeter) An instrument that measures solar radiation, built around a sensor that detects the difference in temperature between two adjacent materials.

pyrgeometer An instrument similar to a pyranometer that measures infrared radiation.

pyrheliometer A very sensitive instrument used to measure the intensity of solar radiation, comprising a blackened surface positioned at right angles to the sunlight, with an extremely accurate thermometer attached to the underside.

quasi-biennial oscillation (QBO) An alternation of easterly and westerly winds in the stratosphere above the Tropics, between about 20° S and 20° N, that happens on average every 27 months.

quasihydrostatic approximation The use of the hydrostatic equation as the vertical equation of motion.

quasi-stationary front A front moving at less than about 5 knots (5.75 mph, 9.25 kmh^{-1}).

QuikScat (Quick Scatterometer) A NASA satellite that carries the SeaWinds radar instrument.

rabal A method used to measure the speed and direction of high-level winds from the elevation and azimuth of a radiosonde balloon.

radar A device used to detect and measure the motion of distant objects by transmitting an electromagnetic wave that is reflected from any object it strikes. Information is obtained by comparing the signal and its reflection. The name is from "*ra*dio *d*etection *a*nd *r*anging."

radar altimetry Measuring the topography of a land surface by means of a radar altimeter.

radar climatology The use of recorded radar echoes showing clouds and precipitation in studies of climate.

radar interferometry A technique used to measure very small changes in the shape of features on the solid surface of the Earth from radar waves that interact with each other to produce characteristic patterns known as interference fringes. These vary in appearance with very small changes in the distance traveled by the radar pulse and its echo.

radar meteorological observation A pattern of echoes on the screen of a weather radar.

radar meteorology The use of radar in compiling a picture of the present state of the weather and preparing weather forecasts.

radar wind A wind observed by radar tracking of a radiosonde balloon or a balloon carrying a radar reflector.

radiant flux density The amount of electromagnetic radiation emanating from a body or falling on a surface.

radiational index of dryness An index calculated from the ratio (R_o/Lr) of the net radiation available to evaporate moisture from a wet surface (R_o) to the amount of heat needed to evaporate the mean annual precipitation (Lr); L is the latent heat of vaporization.

radiation balance The amount of energy the Earth receives from the Sun set against the energy reflected or radiated back into space. It can be summarized as: $R = (Q + q)(1 - a) - I$, where R is the radiation balance, Q is the direct sunlight reaching the surface, q is the diffuse sunlight reaching the surface, a is the surface albedo, and I is the outgoing radiation from the surface.

radiation cooling The loss of energy from the ground at night.

radiation fog Fog that forms on clear nights when the air is moist. At night, the surface radiates heat into the sky and cools sharply. Air adjacent to the ground cools to below the dewpoint temperature, water vapor condenses, and fog forms.

radiation inversion A low-level inversion that forms at night due to radiation cooling.

radiation night A night when the sky is clear.

radiative convective model A one-dimensional climate model that calculates the vigor of convection in a vertical column through the atmosphere.

radiative diffusivity The capacity of an atmospheric layer to distribute infrared radiation by diffusion.

radiative dissipation The loss of energy that occurs when fingers of warm air penetrate cooler air and cool by radiating heat.

radiatus A variety of clouds arranged as broad, parallel bands that appear to converge at a point on the horizon or at two points (radiation points) at opposite sides of the horizon.

radiometer An instrument that measures electromagnetic radiation.

radiosonde A package of instruments carried beneath a balloon that measures atmospheric conditions and transmits the data by radio to a surface station.

raffiche A gusty mountain wind in the Mediterranean region.

rain Liquid precipitation consisting of droplets between 0.02 inch and 0.2 inch (0.5–5.0 mm) in diameter.

rainbow An arch of violet, indigo, blue, green, yellow, orange, and red colored bands, with violet on the inside of the arch and red on the outside. Sometimes a secondary rainbow, with the colors reversed, can be seen about 8° above the primary bow. A rainbow occurs when the Sun is behind the observer, less than 42° above the horizon, and rain is falling in front. The observer sees sunlight reflected from the inside of the rear surface of raindrops and refracted as it enters each raindrop and again as it leaves.

rain cloud Any cloud from which rain or drizzle is likely to fall.

rain day A period of 24 hours, beginning at 0900 Universal Time, during which at least 0.08 inch (0.2 mm) of rain falls.

raindrop A drop of water at least about 0.04 inch (1 mm) and typically 0.08–0.2 inch (2–5 mm) across that falls from a cloud. Raindrops fall at 14–20 mph (23–33 kmh[-1]).

rainfall frequency A measure of how often rain falls in a particular place.

rain gauge The instrument used to measure the amount of rainfall during a given period, usually one day. The internationally approved standard rain gauge is a cylinder 7.9 inches (20 cm) in diameter mounted vertically with its top 39.4 inches (1 meter) above ground level. Rain enters a funnel leading into a measuring tube. In the tipping-bucket

standard gauge   tipping bucket gauge

collecting funnel   heated collecting funnel

2.5
2.0
1.5
1.0
0.5

measuring tube
tipping buckets

rain gauge

gauge, rainwater entering the funnel is guided to a second, smaller funnel and from there into one of two buckets mounted on a rocker. When the bucket is full, its weight tips it downward, making an electrical contact that is transmitted to a recording pen, which moves on a graph mounted on a rotating drum. That bucket empties and the second bucket is positioned to collect water. There are also gauges that automatically record the height of water in the measuring tube and rainfall can be measured automatically by a weighing gauge. An optical rain gauge transmits an infrared beam that is scattered forward when it strikes a raindrop. The detector records the scattering, from which the number of raindrops per second and in a unit volume of air can be counted. This provides a continuous record of rainfall intensity, but it does not measure the amount of rain.

rainmaking Any attempt to induce precipitation to fall from a cloud that otherwise might not have released it.

rainout The removal from the air of solid particles small enough to act as cloud condensation nuclei.

rain shadow The dry region on the lee side of a mountain. Air crossing the mountain loses moisture on the windward side and subsiding air on the lee side warms adiabatically, making it still drier.

rain squall A squall accompanied by a short period of heavy rain.

rainy climate A climate in which the amount of rainfall is adequate to support the growth of plants that are not adapted to dry conditions.

rainy season A period each year when the amount of precipitation is much higher than at other times. This may occur in winter or in summer.

rainy spell A period during which more rain falls than is usual for the place and time of year.

random forecast A weather forecast in which the value for one feature is selected at random.

Rankine (R) A temperature scale in which $1°R = 1°F$, but that extends to absolute zero: $0°R = -459.6°F$. Water freezes at $492°R$ and boils at $672°R$.

Raoult's law When one substance (the solute) is dissolved in another (the solvent), the partial pressure of the solvent vapor in equilibrium with the solution is directly proportional to the ratio of the number of solvent molecules to solute molecules. This means the equilibrium vapor density above the surface of a solution is lower than that above the surface of pure solvent and the more concentrated the solution, the greater the difference. Consequently, more water will enter the solution. Cloud droplets are often quite concentrated solutions. Raoult's law shows that these will grow by the condensation of more water vapor.

rasputitsa A rainy season nearly every year in Russia, lasting several weeks in spring and autumn.

ravine wind A wind that blows along a narrow mountain valley or ravine.

raw An adjective describing weather that is cold, damp, and sometimes windy.

rawinsonde A radiosonde that carries a radar reflector, so it can be tracked to monitor the wind.

Rayleigh atmosphere An idealized atmosphere consisting only of molecules and particles smaller than about one-tenth the wavelength of the solar radiation passing through it.

Rayleigh number A dimensionless number describing the amount of turbulent flow in air being heated from below by convection. $Ra = [(g\Delta\theta)/kv\theta](\Delta z)^3$, where Ra is the Rayleigh number, g is the gravitational acceleration, $\Delta\theta$ is the potential temperature lapse in a layer of air with a depth Δz, θ is the average potential temperature, k is the thermal conductivity of the air, and v is the kinematic viscosity of the air.

Rayleigh scattering The way the direction of solar radiation is altered by its interaction with air molecules and particles. Radiation is scattered most when the molecules and particles are smaller than the wavelengths of radiation.

Réamur temperature scale A temperature scale in which water freezes at 0°R and boils at 80°R; 1°R = 1.25°C = 2.25°F.

reduced pressure The air pressure measured by a barometer at a particular location after it has been corrected to bring it to a sea-level value.

reflection The "bouncing" of light at an angle equal to the angle at which it strikes an opaque surface.

refraction The bending of light as it passes obliquely from one transparent medium to another through which it travels at a different speed. The extent of refraction is proportional to the difference in the speed of light in the two media and the angle at which the light enters.

refractive index The ratio of the speed of light in air to that in another medium. This is a constant for the medium. Air has a refractive index of 1.0003, ice 1.31, water 1.33, and window glass 1.5.

relative humidity (RH) The ratio of the mass of water vapor in a unit mass of dry air to the amount required to produce saturation in that air, written as a percentage.

relative vorticity (ζ) The vorticity about a vertical axis that a mass of fluid possesses by virtue of its own motion relative to the surface of the Earth.

remote sensing Obtaining information about a subject without being in direct physical contact with it.

réseau An abbreviation of *réseau mondiale* (French for "global network"), the name given by the World Meteorological Organization to the weather stations that have been chosen to represent the world climate.

reshabar A strong, dry, northeasterly katabatic wind that blows down the sides of the mountain ranges in southern Kurdistan, straddling northern Iraq and Iran.

residence time The length of time any individual atom, molecule, or particle remains in the air.

resistance hygrometer A hygrometer that measures relative humidity directly, with an accuracy of ±10%, by exploiting the fact that the electrical resistance of certain materials changes with variations in relative humidity.

resonance trough A trough that forms at some distance from a major trough and is related to it.

respiration The process by which living organisms release energy through the oxidation of carbon.

response time The time that elapses between a change in the amount of energy that is available in one part of the climate system and the effect that energy produces.

resultant wind The average speed and direction of the wind at a particular place over a specified period.

return period The frequency with which a rare natural phenomenon may be expected to occur.

return stroke The most visible part of a lightning flash, carrying positive electrical charge toward the cloud from which the stepped leader originated.

revolving storm A storm in which the air moves cyclonically about a low-pressure center.

Reynolds effect A process by which cloud droplets grow in warm clouds. Water evaporates from warmer droplets and condenses onto cooler ones.

Reynolds number *(Re)* A dimensionless number used to measure the extent to which a fluid flows smoothly or turbulently. For air, $Re = LV/v$, where L is the distance the air moves, V is its velocity, and v is its kinematic viscosity (approximately 16×10^{-5} ft.^{2} s^{-1}; 1.5×10^{-5} m^2 s^{-1}). If Re is less than about 1,000, the flow is dominated by viscosity; if greater than about 1,000, by turbulence. Except on a very small scale Re is usually much larger than 10^3.

Richardson number *(Ri)* A mathematical value used to predict whether atmospheric turbulence is likely to increase or decrease. It represents the ratio of the rate at which the kinetic energy of the turbulent motion is being dissipated to the rate at which it is being produced. If Ri is greater than 0.25, turbulence will decrease and disappear. If Ri is less than 0.25, turbulence will increase.

ridge A long, tongue-like protrusion of high pressure into an area of lower pressure.

rime frost A layer of ice that is white and has an irregular surface.

ring vortex The upward spiral around the edge of the vortex in a microburst. It produces a ring of low air pressure.

roaring forties The belt between 40° S and 50° S, noted for its fierce wind storms.

Römer temperature scale A temperature scale in which water freezes at 7.5°, boils at 60°, and average body temperature is 22.5°. The scale is no longer used, but it is the one from which the Fahrenheit scale was derived.

Rossby number *(Ro)* A dimensionless number that is the ratio of the acceleration of moving air due to the pressure gradient and the Coriolis effect: $Ro = U/\Omega L$, where U is the horizontal wind velocity, Ω is the angular velocity of the Earth, and L is the horizontal distance the wind travels.

Rossby waves (long waves, planetary waves) Waves, with wavelengths of 2,485–3,728 miles (4,000–6,000 km), that develop in moving air in the middle and upper troposphere.

rotating-cups anemometer An anemometer consisting of three or four hemispherical or conical cups mounted on arms and attached to a vertical axis. The wind causes the cups to turn about the axis. The rotational speed is converted into wind speed and displayed on a panel below the cups or on a dial.

rotor

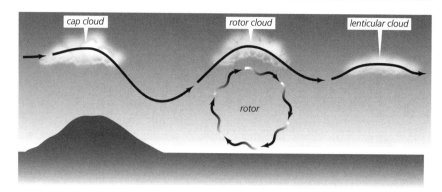

rotor   Rotating air on the lee side of a mountain peak.

rotor cloud   A cloud that forms at the crest of the rotor, as the first of a series of lee waves on the downwind side of a very steep mountain.

roughness length (z_0) The height at which friction with the surface reduces the wind speed to zero.

run of wind   The "length" of a wind, from which its speed can be calculated. It is obtained by multiplying the number of revolutions made over a known period by a rotating-cups anemometer by the circumference of the circle it describes.

rural boundary layer   In a rural area adjacent to a large city, the layer of air between the top of tall vegetation and the uppermost limit of the region within which the climatic properties of the air are modified by the surface.

Saffir/Simpson hurricane scale   A five-point extension to the Beaufort wind scale introduced in 1955 by meteorologists at the U.S. Weather Bureau (now part of the National Weather Service). As well as wind speed and a general description of possible wind damage, it includes the atmospheric pressure at the storm center and the height of the storm surge.

Saharan depression   A depression that forms in winter over the western Mediterranean.

Saharan high   The subtropical high lying permanently over the Sahara Desert.

salt haze   A thin haze caused by the condensation of water vapor onto salt crystals when the relative humidity is below about 90%.

samoon   A hot, dry, northwesterly föhn wind that blows from Kurdistan across Iran.

sand storm A wind storm that lifts sand grains into the air and transports them, often for long distances.

Santa Ana A wind of the föhn type that occurs in southern California, most commonly in autumn and winter.

sastruga A wave in the snow and ice of Antarctica, usually about 2 inches (5 cm) high but in places more than 6 feet (1.8 m) high, that forms where the katabatic winds blow constantly. Sastrugi are aligned parallel to the wind.

saturated adiabat (moist adiabat, wet adiabat) A line on a tephigram that marks the constant wet-bulb potential temperature.

saturated adiabatic lapse rate (SALR) The rate at which rising, saturated air cools adiabatically and descending air warms. It varies according to the air temperature from about 2.7°F per 1,000 feet (5°C km^{-1}) to 5°F per 1,000 feet (9°C km^{-1}). An average value is about 3°F per 1,000 feet (6°C km^{-1}).

saturated adiabatic reference process An idealized representation of the way moist air behaves that is used as a standard, or reference, with which events in the real atmosphere can be compared.

saturated air Air that can hold no more water vapor at its present temperature; its relative humidity is 100%.

saturation The condition in which the moisture content of the air is at a maximum.

saturation deficit (vapor-pressure deficit) The difference between the actual vapor pressure and the saturation vapor pressure at the same temperature.

saturation mixing ratio The value of the mixing ratio of saturated air at a particular temperature and pressure.

saturation ratio The ratio of the actual specific humidity to the specific humidity of saturated air at the same temperature and pressure.

saturation vapor pressure The vapor pressure at which the water vapor in the layer of air immediately above the surface of liquid water is saturated at a given temperature.

scalar quantity A physical quantity that does not act in a particular direction, or for which the direction of action is unimportant or not specified.

scale height The thickness the atmosphere would have if its density were constant throughout at its sea-level value of 1.23 kg m^{-3}. The scale height is 5.2 miles (8.4 km).

scanning multichannel microwave radiometer (SMMR) An instrument carried on the *Seasat* and *Nimbus-7* satellites that measures sea-surface temperature, wind speed, water vapor, clouds and cloud content, snow cover, type of snow, rainfall rates, and different types of ice, including sea ice.

scattered cloud Cloud that covers up to 4 oktas (five-tenths) of the sky.

scattering The change in direction of light as it collides with air molecules and particles. The efficiency of scattering is proportional to the fourth power of the wavelength of the light (λ^{-4}). Particles smaller than the wavelength scatter light less efficiently than bigger particles.

scatterometer A satellite instrument that measures the scattering of radar waves by the small capillary waves on the ocean surface, from which the speed and direction of the surface wind can be calculated.

scavenging The removal of particles from the air by precipitation.

Scharnitzer A cold, persistent northerly wind in the Tirol region, Austria.

scotch mist Stratus cloud that forms suddenly on high ground and is very common on hills and mountains in Scotland and in other high-latitude regions.

screened pan An evaporation pan covered by a wire mesh screen that reduces insolation and evaporation, producing a pan coefficient closer to unity than that from an unscreened pan.

scud Fragments of tattered cloud, most commonly nimbostratus, lying below the general cloud base.

sea-breeze front The boundary formed when cool, maritime air advances beneath warmer air over land, producing a sea breeze.

sea fog Advection fog that forms at sea when warm, moist air moves over an area of much colder water.

sea ice Ice that forms by the freezing of sea water; sea water with a salinity of 35 at standard sea-level pressure freezes at 28.56°F (–1.91°C).

sea-level pressure The atmospheric pressure at sea level. This varies, but the average value used in the definition of the standard atmosphere is 101.325 kPa (= 1,013.25 mb, 760 mm mercury, 29.9 inches of mercury, 14.7 lb. per square inch).

Seasat The first satellite to use imaging radar to study the Earth. On October 10, 1978, after transmitting data for 106 days, a short circuit drained all the power from its batteries and *Seasat* ceased to function.

seasonal drought A drought that occurs in climates where all or most of the precipitation falls during one season.

sea-surface temperature (SST) The temperature of the water at the surface of the sea, routinely measured by drifting buoys, ships, and orbiting satellites.

sea turn A name in the northeastern United States, and especially New England, for a wind from the sea, often bringing mist.

Sea Winds A radar instrument on the *QuikScat* satellite that gathers data on low-level wind speed and direction over the oceans and tracks the movement of antarctic icebergs.

seca A severe drought or very dry wind in northeastern Brazil.

seclusion An occlusion in which the cold front first starts to overtake the warm front some distance from the peak of the wave depression.

secondary air mass An air mass that has been modified by passing over a surface different from the one over which it developed its characteristics.

secondary circulation That part of the general circulation of the atmosphere that consists of relatively small-scale, short-lived features superimposed on the primary circulation.

secondary cold front A front that develops in the cold air behind a frontal cyclone when the horizontal temperature gradient is so strong that the cold air starts to separate into two distinct masses, one colder than the other.

secondary cyclogenesis The development of a second extratropical cyclone as the first cyclone becomes occluded and starts filling.

secondary cyclone A small cyclone that forms within the circulation of a larger (primary) cyclone.

secondary front One of the fronts that form along a frontal wave behind the first (primary) front.

secondary pollutant A pollutant produced in the atmosphere by chemical reactions between primary pollutants.

sector A horizontal plane bounded on two sides by the radii of a circle and on the third by an arc forming part of the circumference of the same circle.

sector wind The average direction and speed of the wind at the altitude at which an aircraft plans to fly over one sector of the route it intends to follow.

sector

Seebeck effect If two wires or rods of dissimilar materials are welded together at both their ends, and one of the joints is at a different temperature from the other, an electric current will flow through the circuit.

seistan An almost incessant north or northwest wind between June and September in eastern Iran and western Afghanistan, but mainly in the Iranian region of Seistan.

selatan A strong, dry southerly wind over parts of Indonesia during the southeast monsoon.

semidiurnal Occurring twice in every 24 hours.

sensible temperature The temperature the body feels. This is not always the same as the temperature measured by a thermometer, because of such factors as wind chill and high humidity.

settled An adjective describing fine weather conditions that remain unchanged for a minimum of several days and more commonly for a week or more.

severe storm Any storm that damages property or endangers life. The U.S. National Weather Service defines a severe thunderstorm as one producing hailstones 0.75 inch (19 mm) or more across, wind gusts of 58 mph (93 kmh^{-1}) or more, a tornado, or more than one of these.

severe-storm observation A report of a severe storm that has been positively identified.

sferics A word derived from "atmo*spherics,*" to describe the electromagnetic disturbances caused by natural electrical phenomena. A sferics receiver uses two square radio aerials mounted at right angles to each other and can detect the direction of a thunderstorm up to a distance of about 1,000 miles (1,600 km) away. Two or more sferics receivers can pinpoint the location of a thunderstorm.

shade temperature The air temperature measured anywhere out of direct sunlight. A thermometer exposed to sunlight will absorb solar radiation directly. This will raise its temperature above that of the surrounding air and consequently it will give a false reading.

shallow fog Fog that extends no more than 6 feet (1.8 m) above the surface.

shamal A hot, dry wind that blows almost continuously during June and July over Iraq, Iran, and the Arabian Peninsula.

sharp-edged gust A gust of wind that changes the speed or direction of the wind almost instantaneously.

shear A force that acts parallel to a plane, rather than at right angles to it.

shear instability (Kelvin–Helmholtz instability) The situation in which two layers of a fluid are moving in the same direction but at different speeds.

shear line A line or narrow belt marking an abrupt change in the wind direction or speed.

shear wave A wave that forms where there is strong horizontal wind shear in stable air. The difference in wind speed between two layers of air causes turbulence. Air from the lower layer rises, but its stability causes it to sink again, establishing a wave pattern.

sheet lightning Lightning seen as a general flash, rather than being precisely located. It may be forked lightning between two clouds, seen through an intervening cloud, or a lightning flash between the separated charges inside a cloud.

shelf cloud A layer of cloud that projects like a shelf beneath the anvil of a big cumulonimbus storm cloud.

shelter belt A line of trees grown at right angles to the prevailing wind in order to reduce the wind speed on the lee side, thereby protecting ground or crops that might be damaged by strong winds.

shelter temperature (air temperature, surface temperature) The temperature registered by a thermometer about 4 feet (1.25 m) above the ground inside a Stevenson screen or other shelter. This is the minimum height at which a thermometer gives an accurate reading, free from surface effects.

shielding layer A name sometimes given to the planetary boundary layer, which shields the surface from events in the free atmosphere.

shimmer The appearance of shimmering in the air immediately above a hot road surface. It is a mirage, caused by light refraction, in the form of an inferior image of the sky.

ship code A version of the international synoptic code that is approved by the World Meteorological Organization for use by voluntary observing ships.

ship report A weather report compiled on board a voluntary observing ship at sea and transmitted to a shore receiving station.

short-range forecast A weather forecast that covers a period of up to two days.

shower A short period of precipitation from a convective cloud that starts and ends abruptly, varies in intensity but can be heavy, and is often followed by sunshine and a blue sky.

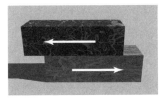

shear

shrieking sixties The region between latitude 60° S and the edge of the ice surrounding Antarctica, notorious for its fierce storms.

Shūrin season A period of high rainfall that occurs in Japan during September and early October.

Siberian high A region of high surface pressure that forms over Siberia in winter, centered to the south of Lake Baikal.

siffanto A southwesterly wind, often violent, that blows over the Adriatic Sea and surrounding lands.

sigua A gale that blows in the Philippines.

silver iodide A yellow, solid compound of silver (AgI) that is used in cloud seeding because it can be made to form particles the size of freezing nuclei.

silver thaw The name given to freezing rain in the Pacific Northwest. It is called "thaw" because it is often followed by warmer weather.

simoom A hot, dry, usually dusty wind of the sirocco type that blows in spring and summer across the southeastern Sahara and the Arabian Peninsula.

singular corresponding point A center of high or low pressure shown on a constant-height or constant-pressure chart that reappears on succeeding charts.

singularities Certain types of weather that occur fairly regularly at a particular time each year, e.g., the Indian summer, anticyclones at the end of June, and the January thaw around January 20–23 in the northeastern United States.

sirocco A hot wind that blows across countries bordering the Mediterranean, most commonly in spring.

skill score A measure of the accuracy of a weather forecast, made by comparing the forecast with a reference standard.

sky cover The proportion of the sky that is obscured by clouds, expressed in oktas or tenths.

sky map A pattern of light and dark areas sometimes seen on the underside of a cloud layer, most commonly in high latitudes when the surface is partly covered by snow or ice.

sky-view factor The amount of sky that can be seen from a particular point on the surface, expressed as a proportion of the total sky hemisphere.

sleet (1) Small raindrops that freeze into ice particles as they are falling. (2) In Britain, a mixture of rain and snow falling together.

slice method A method used to study static stability in a layer of air by taking account of air moving vertically through the "slice" in both directions.

sling psychrometer An alternative name for a whirling psychrometer.

slush A mixture of melting snow and/or ice and liquid water lying on the ground.

small hail Hail consisting of hailstones less than about 0.2 inch (5 mm) in diameter (smaller than garden peas), but sufficiently solid to remain intact until they hit the ground. They are too small to cause any damage or injury.

small ion A charged particle of dust or other substance that exists as an aerosol.

small-ion combination The removal of small ions through their reaction with other particles.

smog (1) A mixture of smoke and fog. (2) Photochemical smog; a haze produced in strong sunlight by chemical reactions between pollutants.

smoke An aerosol produced by the incomplete combustion of a carbon-based fuel that consists of solid or liquid particles, most less than 0.00004 inch (1μm) in diameter.

smoke horizon The top of a layer of smoke that is trapped beneath an inversion and seen from above against a clear sky. The boundary between the smoke and clear air forms a false horizon.

smokes Clouds of dust and dense, white haze and mist that form in the morning and evening during the dry season near the western coast of equatorial Africa.

smudging Using oil-burning heaters (smudge pots) to protect a delicate farm crop against frost damage. Their smoke forms a layer above the ground, reducing the loss of heat by infrared radiation from the surface.

snap A short period of unusually cold weather that commences suddenly.

snow Precipitation that falls as aggregations of ice crystals.

snow accumulation The depth of snow lying on the ground at a particular time.

snowball Earth The Earth when its entire surface, apart from the highest mountains, is covered by ice. Some scientists believe this occurred four times between 750 million and 580 million years ago. Mean temperatures were about –58°F (–50°C) and all the oceans were frozen to a depth of more than 0.6 mile (1 km).

snow banner (snow plume, snow smoke) The appearance, when seen from a distance, of snow that is being blown from a mountain top or other exposed high ground.

snow belt The strip of land, up to about 50 miles (80 km) wide, parallel to the lee shore of a large lake where winter snowfall is markedly higher than it is on the upwind side of the lake.

snow blink (snow sky) The bright, glaring appearance of the underside of clouds when the ground is covered with snow.

snowblitz A theory that a series of processes might reinforce each other and cause the Northern Hemisphere to be plunged into an ice age in a matter of decades.

snow chill The effect of being covered in snow that melts, taking latent heat from the body and saturating the fibers of clothing with cold water.

snow cloud A cloud from which snow falls or appears likely to fall.

snow cover All the snow lying on the ground, or the proportion of an area covered by snow. Ground is usually said to be snow-covered if more than 50% is beneath snow.

snow-cover chart A synoptic chart showing the areas covered by snow and with contour lines showing the depth of snow.

snowdrift An accumulation of snow that is much deeper than the snow covering adjacent areas. Drifts occur where moving air has lost a significant amount of its energy by friction with the surface and therefore its ability to keep snowflakes aloft.

snow dune A snowdrift shaped like a sand dune, having formed in the same way.

snow dust Very fine, powdery snow driven by the wind.

snow eater A warm, dry wind that removes snow by sublimation. A chinook wind often removes 6 inches (15 cm) of snow in a day and can clear 20 inches (50 cm).

snowfall Precipitation that falls as snow. Because snow is more bulky than rain but variable in composition, amounts of snowfall are converted to their rainfall equivalents when reporting precipitation amounts. At about freezing, 10 inches of snow is equivalent to 1 inch of rain.

snow field An extensive, approximately level area that is covered uniformly with fairly smooth snow or ice.

snowflake An aggregation of ice crystals grouped into a regular six-sided or six-pointed shape about 0.04–0.8 inch (1–20 mm) across.

snow flurry A light snow shower of brief duration.

snow gauge A modified rain gauge used to measure the amount of snow that has fallen over a stated period.

snow geyser Fine, powdery snow that is suddenly thrown upward when a thick layer of underlying snow settles abruptly.

snow grains (granular snow, graupel) Very small, flat particles of white, opaque ice, less than 0.04 inch (1 mm) in diameter.

snow line (1) The boundary between an area covered by snow and the area free from snow. (2) The edge of the snow that remains on a mountainside throughout the summer.

snowout The removal from the air of solid particles that act as freezing nuclei or cloud condensation nuclei and are carried to the ground in snow.

snow pellets (graupel, soft hail) Spherical or less commonly conical, opaque, white grains of ice 0.1–0.2 inch (2–5 mm) in diameter.

snow stage The point at which water vapor in rising air changes directly into ice by sublimation. This happens when the condensation level is higher than the freezing level.

snow storm A heavy fall of snow; there is no precise definition of the term.

soft UV (UV-B) Ultraviolet radiation at wavelengths of 280–315 nm.

solaire An easterly wind in central and southern France, referring to the fact that easterly winds blow from the direction of the rising Sun (in French, *soleil*).

solar air mass The total optical mass of the atmosphere that is penetrated by sunlight with the Sun at any given position in the sky.

solar constant The amount of energy the Earth receives from the Sun per unit area (usually per square meter) calculated at a point perpendicular to the Sun's rays and located at the outermost edge of the Earth's atmosphere. The value of the solar constant is not known precisely, but the best estimate is 1,367 W m^{-2} (1.98 langleys).

solar flare A sudden increase in the strength of the solar wind associated with sunspot activity. A major flare can double the amount of energy emitted by the Sun over a period of about 0.25 second.

solar irradiance The total amount of energy the Sun emits. The total amount of energy radiated by the Sun is about 70 MW m^{-2} and over the entire surface of the photosphere about 4.2×10^{20} MW.

solar radiation cycle The regular increase and decrease in the amount of radiation emitted by the Sun over an 11-year cycle.

solar–topographic theory A theory that explains past changes in climate in terms of variations in solar output and the formation and erosion of mountains.

solar wind A stream of protons, electrons, and some nuclei of elements heavier than hydrogen that flows outward from the Sun.

solstice (midsummer day and midwinter day) One of the two dates each year when the difference in length between the hours of daylight and darkness is most extreme. At present these fall on June 21–22 and December 22–23.

solute effect The effect whereby water vapor condenses more readily onto cloud droplets that condensed onto hygroscopic nuclei than onto droplets of pure water. This is because hygroscopic nuclei dissolve into the droplets, forming solutions, and the saturation vapor pressure is lower over the surface of a solution than over a surface of pure water.

sonde A package of instruments carried by a free-flying balloon that measures temperature, pressure, and humidity in the air through which they move. When the balloon bursts, the instruments parachute to the ground.

sonic anemometer An anemometer that calculates wind speed from its effect on the speed of a sound signal.

sounder An instrument that makes a measurement through a column of air.

sounding Any measurement that is made through the column of air between the instrument and the level that is being monitored.

source A place at which a particular substance, usually a pollutant, is released into the environment, or the process by which it is released.

source region An extensive area within which the surface is fairly uniform, pressure systems are stationary for most of the time, and consequently where a particular type of air mass originates.

southerly burster A strong, cold, southerly wind that blows across the eastern part of New South Wales, Australia.

southern circuit The path, sometimes leading as far south as the Gulf of Mexico, that depressions usually follow as they cross the United States from west to east during winter.

southern oscillation A change in the distribution of air pressure over the equatorial South Pacific that occurs at intervals of one to five years and is associated with El Niño and La Niña events.

southern oscillation index (SOI) A measure of the difference in sea-level atmospheric pressure between two monitoring stations, those most often used being at Darwin, Australia, and Tahiti, in the central South Pacific.

south temperate zone That part of the Earth between the tropic of Capricorn and Antarctic Circle.

spatial dendrite An approximately spherical ice crystal with branches extending in all directions from a central nucleus.

special observation An observation of a particular weather condition affecting the operation of aircraft that is made because of a significant change since the most recent report.

special sensor microwave imager (SSM/I) A passive microwave radiometer carried on satellites of the Defense Meteorological Satellite Program that provides data on clouds and other meteorological phenomena.

specific gas constant *(R)* A value used when the gas laws are combined into the equation of state. $R = 10^3 R*/M$, where $R*$ is the universal gas constant (measured in joules per kelvin per mole) and M is the molecular weight of the gas (the value must be multiplied by 1,000 because moles are defined in grams and the unit of mass is the kilogram).

specific humidity The ratio of the mass of water vapor present in the air to a unit mass of that air including the water vapor.

specific volume The volume occupied by a unit mass of a substance.

spectrum of turbulence The range of frequencies of the oscillations making up turbulent flow.

spell of weather A period, usually of five to 10 days, during which particular weather conditions persist.

spillover Precipitation due to orographic lifting that is blown over the top of a mountain by the wind and falls inside what is ordinarily the rain shadow.

spiral band A pattern on a radar screen made by echoes from the center of a tropical cyclone.

spissatus (spi) A species of cirrus cloud that is sufficiently dense to appear grayish when viewed looking toward the Sun.

spontaneous nucleation The formation of water droplets or ice crystals in the absence of cloud condensation nuclei or freezing nuclei. Water vapor will condense into liquid spontaneously when the relative humidity exceeds about 101%, and ice crystals will form spontaneously when the temperature falls below –40°F (–40°C).

Spörer minimum The period from 1400 to 1510, during which very few sunspots were observed. Temperatures were abnormally low and, like the Maunder minimum, the period was known as the Little Ice Age.

spot wind The wind that is observed or forecast at a specified height over a specified location.

squall A sudden, brief storm in which the wind speed increases by up to 50% to at least 16 knots (18.4 mph, 30 kmh^{-1}) for at least two minutes, then dies away more slowly.

squall cloud A roll of dark cloud, often with mammatus, that forms along the leading edge of a squall line.

squall line A series of very vigorous cumulonimbus clouds that merge to form a continuous line, often up to 600 miles (965 km) long, which advances at right angles to the line itself.

stability chart A synoptic chart showing the distribution of values given by a particular stability index.

stability index One of a series of values used to summarize the stability of air and the severity of the thunderstorms, up to and including tornadoes, likely to be generated by varying degrees of instability.

stack height (chimney height) The actual height above the ground of the top of a factory smoke stack (chimney).

stade (stadial) A cold period that occurs during a glacial period.

standard atmosphere (standard pressure) The average atmospheric pressure at mean sea level, assuming the atmosphere to consist of a perfect gas at a temperature of 59°F (15°C, 188.16K) and the acceleration due to gravity to be 9.80655 m s^{-2}. The standard atmosphere is defined as a pressure of 1.013250×10^5 newtons per square meter (= 760 millimeters [29.9213 inches] of mercury [density 13,595 kilograms per cubic meter], 14.691 pounds per square inch, or 1,013.25 millibars).

standing cloud A stationary cloud, usually above or close to a mountain peak or other high ground.

state of the sky A full description of the sky, including the amount, type, and height of all clouds and the direction in which they are moving.

static stability (convective stability) The condition in which the atmosphere is stratified so that its density decreases and potential temperature and buoyancy increase with height.

stationary front A front at which the air on either side is moving approximately parallel to the front. The surface position of the front does not move, or moves erratically and slowly, and may remain in the same position for several days. A stationary front is shown on weather maps as a line with semicircles on one side and triangles on the other.

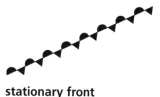

stationary front

station elevation The vertical distance between a weather station and mean sea level.

station model The formalized diagram used to report observations from a weather station. Standard symbols represent cloud, precipitation, and wind direction and speed. Other information is given in numbers. Each item of information occupies a particular position around the station, which is represented by a central station circle.

station pressure (surface pressure) The atmospheric pressure measured at the station elevation.

statistical forecast A weather forecast compiled from studies of past weather patterns that are compared with present conditions and used to assess the statistical probability of the pattern repeating.

statistically dynamical model A two-dimensional climate model that deals only with surface processes. It describes the latitudinal transport of energy and vertical distribution of energy by radiation and convection.

steam devil A stream of fog that forms an almost vertical column of cloud over the surface of an unfrozen lake in winter.

steam fog Thin, wispy fog that forms when cold air crosses a warmer water surface.

Stefan–Boltzmann constant (σ) The amount of radiant energy released by a black body; $\sigma = 5.67 \times 10^{-8}$ W m^{-2} K^{-4} (watts per square meter per kelvin to the fourth power).

Stefan–Boltzmann law The physical law relating the amount of radiant energy a body emits to its temperature. It is expressed as: $E = \sigma T^4$, where E is the amount of radiation emitted in watts per square meter,

hygrometer

maximum and minimum thermometers

Stevenson screen

T is the temperature in kelvins, and σ is the Stefan–Boltzmann constant.

stem flow Precipitation that falls onto vegetation and reaches the ground by running down the stems of plants.

stepped leader The first stage in a lightning flash, which carries negative charge away from the cloud as a stream of electrons.

Stevenson screen The container housing the thermometers and hygrometers or hygrographs used at a weather station. It is a box, painted white and with louvered walls of double thickness on all sides. It stands on legs that raise it so that the bulbs of its thermometers are about 4 feet (1.25 m) above the ground.

stochastic resonance The observable effect that results when a stochastic process acts in the same sense as a natural cycle that is too weak to produce any effect by itself.

Stokes's law A law describing the factors that determine the friction experienced by a spherical body falling by gravity through a viscous medium. Friction *(F)* is given by: $F = 6\pi r \eta v$, where r is the radius of the body, v is its velocity, and η is the viscosity of the medium.

storm A wind of force 11 on the Beaufort wind scale, which blows at 64–75 mph (103–121 kmh^{-1}).

storm center The area of lowest surface atmospheric pressure in a cyclone.

storm glass An instrument indicating changes in the weather that was popular in the 18th and 19th centuries. It consisted of a heavy, tightly sealed glass tube containing a supersaturated mixture of chemical compounds. Crystals would form and dissolve inside the glass, the changes apparently being linked to meteorological changes. Clear liquid meant the weather would be fine. Cloudy liquid meant it would rain. If crystals formed in winter at the bottom of the glass, there would be frost.

storm model A three-dimensional simulation of the way air and water vapor move into, out of, and vertically within a storm.

storm surge A rise in sea level, accompanied by huge waves, that is produced by large storms at sea and especially by tropical cyclones.

stowed wind A wind that is partly blocked by a physical barrier, such as a range of mountains or hills, so it is forced through gaps between them.

Strahler climate classification A climate classification based on the general circulation of the atmosphere. World climates are grouped into three main types according to the air masses and prevailing winds that control them. These are subdivided further into 14 climatic regions, plus highland climates.

stratiform An adjective describing clouds that form extensive horizontal layers.

stratiformis (str) A species of cumuliform clouds that spread across the sky to form an extensive sheet.

stratocumulus (Sc) A genus of low clouds seen as patches, sheets, or layers of gray, white, or both gray and white cloud. There are always dark rolls or rounded masses. These sometimes merge into larger masses. The smallest elements have an apparent width of about 5°.

stratopause The boundary, where temperature ceases to change with height, that separates the stratosphere from the mesosphere. In summer, it is at a height of about 34 miles (55 km) over the equator and Poles and about 31 miles (50 km) in middle latitudes. In winter, it is at about 30 miles (48 km) over the equator and 37 miles (60 km) over the Poles. The atmospheric pressure at the stratopause is about 1 mb.

stratosphere The layer of the atmosphere above the tropopause, extending to an altitude of about 31 miles (50 km).

Stratospheric Aerosol and Gas Experiment (SAGE) A set of instruments carried on the *Earth Radiation Budget Satellite* and used to measure the material injected into the stratosphere by volcanoes.

stratospheric coupling Any interaction between disturbances on either side of the tropopause.

stratospheric wind Wind that blows in the stratosphere.

stratus (St) A genus of low clouds that form a uniform, gray layer. When stratus forms at the surface, it is fog.

streak cloud An elongated fragment of fibrous cloud, commonly cirrus, that indicates the direction and strength of the wind shear.

streak lightning A lightning stroke in which the main channel has branches. It may flash between a cloud and the ground or between a cloud and adjacent air.

streamline The track followed by moving air. A wind streamline is shown on a map or diagram as a line parallel to the wind direction at every point along its path.

strong breeze A force 6 wind on the Beaufort wind scale, which blows at 25–31 mph (40–50 kmh^{-1}).

streamline

strong gale A force 9 wind on the Beaufort wind scale, which blows at 47–54 mph (76–87 kmh^{-1}).

Stüve chart (adiabatic chart, pseudoadiabatic chart) A thermodynamic diagram in which temperature is plotted along the horizontal axis and pressure along the vertical axis, with the highest pressure at the bottom.

subcloud layer Some or all of the air below the cloud base.

subgeostrophic An adjective describing a wind with less force than the geostrophic wind.

subgradient wind A wind with less force than the gradient wind indicated by the pressure gradient and latitude (which determines the magnitude of the Coriolis effect).

sublimation The change of ice into water vapor, without passing through a liquid phase.

sublimation nucleus A solid particle in the air onto which water vapor solidifies by deposition (sometimes called sublimation) to form an ice crystal.

subpolar low A belt of low atmospheric pressure between latitudes 60° and 70° in both hemispheres, where the polar easterly and midlatitude westerly winds converge.

subpolar region The part of the world between the low-latitude margin of land occupied by tundra and the high-latitude margin of lands with cool temperate or desert vegetation.

subsidence A general sinking of air over a large surface area, producing high surface air pressure and divergence.

substratosphere The uppermost part of the troposphere, immediately below the tropopause.

subsun A bright spot of light seen from above on the top of a cloud layer.

subsynoptic An adjective describing conditions over an area that is large, but not so large as in the picture presented in a synoptic view.

subtropical cyclone A cyclone that develops in the subtropics when the southern tip of a polar trough in the upper atmosphere becomes cut off from the main part of the trough.

subtropical easterlies index A value that is calculated for the strength of the easterly winds in the subtropics between latitudes 20° N and 35° N.

subtropical high One of several semipermanent anticyclones located over the ocean in the subtropics. They are most developed in summer and strongly influence climates to the east of them by blocking or diverting depressions traveling from west to east in middle latitudes.

subtropical high-pressure belt One of the two regions of generally high atmospheric pressure that surround the Earth in the subtropics, centered at about latitude 30° N and 30° S.

subtropical jet streams The jet streams located at about 30° N and S throughout the year.

subtropics The belts in both hemispheres that lie between the Tropics and approximately latitude 35–40°.

suction scar An approximately circular mark left on open ground by a suction vortex.

suction vortex A small column of spinning air that moves in a circle around the main vortex of a tornado. A major tornado may generate two or more suction vortices.

sudestada A strong southeasterly wind that blows in winter along the Atlantic coasts of Argentina, Uruguay, and southern Brazil.

suhaili A strong southwesterly wind that blows across the Persian Gulf in the wake of a depression.

sukhovei A hot, dry, easterly wind that blows across southern Russia and the European steppe, most often in summer.

Sullivan winter storm scale A scale for classifying winter storms that ranks storms as: (1) minor inconvenience; (2) inconvenience; (3) significant inconvenience; (4) potentially life-threatening; and (5) life-threatening.

sultry An adjective describing the uncomfortable conditions that result when a high air temperature coincides with high relative humidity and still air.

sumatra A strong squall that crosses the Malacca Strait during the southwest monsoon.

summation principle A method for reporting the sky cover at any specified level, based on the principle that this is equal to the total sky cover at each cloud layer from the lowest to the level specified. At no level can the cloud cover be less than the cover at lower levels, and the total cloud cover cannot exceed 8 oktas (10/10).

sun cross A rare optical phenomenon comprising a vertical band of white light in the sky with a horizontal band across it.

sun dog (mock sun, parhelion) A bright patch seen to one side of the Sun and usually slightly below it. It is caused by the refraction of sunlight by ice crystals that are falling slowly.

sun photometer An instrument that is held in the hand and pointed directly at the Sun to measure the intensity of direct sunlight.

sun pillar An optical phenomenon in which a bright shaft of light extends upward from the Sun or, much more rarely, below the Sun.

sunshine recorder An instrument that measures the intensity and duration of insolation.

sunspot A dark patch on the visible surface of the Sun where the temperature is lower than the surrounding area. The number of sunspots increases and decreases over an approximately 11-year cycle that coincides with climatic changes on Earth.

sun-synchronous orbit A satellite orbit in which the satellite remains in the same position relative to the Sun, at a height of about 560 miles (900 km). The satellite crosses the equator about 15 times each day and passes over every part of the surface of the Earth at the same time each day.

superadiabatic The condition in which the environmental lapse rate (ELR) is greater than the dry adiabatic lapse rate (DALR).

supercell The type of convection cell that sometimes develops in a very massive cumulonimbus storm cloud. Upcurrents rise at an angle to the vertical, so instead of falling directly into the rising air, the precipitation falls to the side of it and the upcurrents are not suppressed, allowing the cell to continue growing. A supercell storm is capable of producing tornadoes.

supercooling The chilling of water to below its freezing temperature without triggering the formation of ice.

supergeostrophic An adjective describing a wind with greater force than the geostrophic wind.

supergradient wind A wind that is stronger than the gradient wind predicted by the balance of the pressure gradient force and Coriolis effect.

superior image A mirage in which an object appears higher than its real position.

supernumerary bow A rainbow that appears inside the primary bow with the colors in the same order as those in the primary.

super outbreak A chain of 148 tornadoes that occurred on April 3 and 4, 1974, along three squall lines extending from the southern shore of Lake Michigan to Alabama and at one point from Canada to the Gulf of Mexico.

supersaturation The condition of air in which the relative humidity exceeds 100%.

supertyphoon A tropical cyclone in the Pacific Ocean that covers an area up to 3 million square miles (8 million km^2).

surazo A cold, dry, southerly or southwesterly wind that blows across the mountain ranges and high plateau of Peru, often with great force.

surface analysis The study of a surface synoptic chart in order to identify air masses, frontal systems, and other features and to plot their locations.

surface inversion (ground inversion) A temperature inversion that begins at ground level.

surface pressure chart A synoptic chart showing the distribution of station pressure.

surface shearing stress The force exerted on a surface by air that passes across it.

surface temperature The temperature of the air or sea measured close to the surface of land or water.

surface tension A property of a liquid that makes it seem that a thin, flexible skin covers its surface. It is caused by the mutual attraction of molecules in the liquid.

surface visibility The horizontal visibility measured by an observer on the ground.

surface weather observation An observation or measurement of weather conditions made at ground level or on the surface of the sea.

surface wind The speed and direction of the wind blowing at the surface of land or sea.

surge line A line just ahead of a local group of thunderstorms where there is a sudden change in the wind speed and direction.

surge of the trades An acceleration of the trade winds due to changes in the distribution of pressure that increase the pressure gradient away from the centers of the subtropical highs.

swamp model A model of the general circulation in which the oceans are treated as though they were permanently wet land (a swamp).

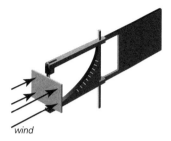

**swinging-plate
anemometer**

sweeping A mechanism by which raindrops are believed to grow by colliding and merging with smaller drops, and other small droplets are swept into the wake of the bigger drop and absorbed by them in that way.

swinging-plate anemometer (pressure-plate anemometer) An anemometer comprising a flat plate that swings freely at the end of a horizontal arm that is free to rotate about a vertical axis. A wind vane fixed to the other end of the arm ensures that the plate is always at right angles to the wind direction. Air pressure makes the plate swing inward and the distance it moves is converted into a wind speed read from a scale.

synoptic An adjective describing something based on a general view over a large area at a particular time.

synoptic chart A map showing a general picture of the weather conditions over a large area at a particular time.

synoptic climatology The branch of climatology dealing with the influence of atmospheric circulation patterns on regional climates.

synoptic code One of the recognized codes used to transmit synoptic weather data and observations.

synoptic meteorology The study of weather conditions over a large area simultaneously.

synoptic report Any weather report based on synoptic weather observations that is encoded using an authorized code and transmitted to a weather center.

synoptic scale (cyclonic scale) The scale of weather phenomena that can be shown on a synoptic chart. These events extend horizontally for about 600 miles (1,000 km), vertically for about 6 miles (10 km), and they last for about one day. These are the approximate dimensions of a cyclone.

synoptic weather observation A set of observations and measurements of surface weather conditions made at a weather station using standard instruments, calibrations, and methods at one of the times specified by the World Meteorological Organization.

synthetic aperture radar (SAR) A type of radar carried on an aircraft or satellite that transmits a continuous-wave signal at a precisely controlled frequency and stores the reflected signals in a memory. The effect is similar to that of an instrument with a large antenna, although only a small antenna is used. SAR is used to map the surface of the Earth or other planets in great detail.

taino   A local name for a tropical cyclone in some parts of the Greater Antilles.

tarantata   A strong northwesterly breeze in the Mediterranean region.

tehuantepecer   A cold, northerly wind that blows almost constantly in winter over the isthmus of Tehuantepec, Mexico.

teleconnections   Linked atmospheric changes that occur in widely separated parts of the world.

Television and Infrared Observation Satellite (TIROS)   The first weather satellite, launched by the United States on April 1, 1960. TIROS satellites are now known as NOAA-class satellites. They travel in polar orbits, scanning the entire surface of the Earth over a 24-hour period. Their instruments are sensitive to visible light and infrared radiation; as well as monitoring the weather, the satellites carry search and rescue transponders.

temperate belt   The regions lying approximately between latitudes 25° and 50° in both hemispheres.

temperate zone   The regions between the Tropics (23.5° N and S) and the Arctic and Antarctic Circles (66.5° N and S).

temperature belt   The area between two lines on a graph that show the daily maximum and minimum temperatures for a particular place.

temperature gradient   The rate of temperature change over a horizontal distance.

temperature–humidity index (comfort index, discomfort index, heat index, THI)   A numerical value relating the temperature and humidity of the air to the conditions a sedentary person wearing ordinary indoor clothes finds comfortable. $THI = 0.4(T_a + T_w) + 15$, where T_a and T_w are the dry-bulb and wet-bulb temperatures respectively, measured in °F. If the temperatures are measured in °C: $THI = 0.4(T_a + T_w) + 4.8$.

temperature range   The difference between the highest and lowest temperatures recorded for a particular place. Annual range uses only daytime temperatures. Diurnal range compares the mean daytime and mean nighttime temperatures for a month, season, or year. Mean range uses only mean temperatures. Absolute range uses the highest and lowest temperatures by day or night.

temporale   A strong southwesterly or westerly wind that blows onto the Pacific coast of Central America in summer.

tendency   The rate of change of a vector quantity at a specified place and time.

tendency interval The period (usually three hours) between the measurements used to determine the tendency of a meteorological factor.

tephigram (TΦgram) A thermodynamic diagram on which the temperature and humidity of the air are plotted against pressure. This reveals the entire structure of a column of air.

terdiurnal Happening every three days.

terminal velocity (fall speed) The maximum speed a falling body can attain. In the lower troposphere, the terminal velocity *(V)* of a body the size of a small raindrop is $8 \times 10^3 r$ meters per second or feet per second ($1\text{m s}^{-1} = 3.3$ ft. s^{-1}), where r is the radius of the body in meters or feet. For large raindrops and hailstones, $V = 250 r^{1/2}$. More precisely, $V = 2\,g\,(\rho_p - \rho_a)\,r^2 \div 9\,\eta$, where g is the acceleration due to gravity (32.18 ft. s^{-2} = 9.807 m s^{-2}), ρ_p is the density of the body, ρ_a is the density, and η the viscosity of the medium through which it is falling.

terral A land breeze that blows with great regularity along the western coasts of Chile and Peru.

terrestrial radiation The black body radiation emitted by the surface of the Earth. This is approximately equal to the radiation emitted by a black body at a temperature of 255K (–0.67°F, –18.15°C).

tertiary circulation The circulation of air on a very local scale, such as a thunderstorm, tornado, or local wind.

thaw A spell of warm weather in winter or early spring during which snow and ice melt.

thawing index The cumulative number of degree days when the air temperature is above freezing.

thawing season The time between the lowest and succeeding highest points on the time curve of cumulative degree days when the air temperature is above and below freezing.

thermal A convection current that forms locally, where the ground has been heated strongly by the Sun.

thermal belt An area on many mountainsides in middle latitudes where nighttime temperatures are higher than the temperature at higher and lower elevations. Cold air subsides at night to produce low temperatures below the thermal belt and the adiabatic decrease in temperature with height produces cold air above the thermal belt.

thermal climate A climate defined only by its temperature.

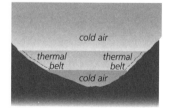

thermal belt

thermal conductivity The rate at which heat passes through a substance, measured as the amount of heat transmitted in unit time over a unit distance in a direction perpendicular to a surface of unit area, under conditions in which the transfer of heat depends only on the temperature gradient.

thermal diffusivity (κ) The rate at which heat penetrates a material. This depends on the thermal conductivity of the material, its density, and its specific heat capacity; $\kappa = k/\rho C$, where k is the thermal conductivity, ρ the density, and C the specific heat capacity.

thermal efficiency index A value for the amount of energy, as heat, that is available for plant growth in the course of a year. The thermal efficiency is $(t - 32)/4$ for each month, and the thermal efficiency index is the sum of the thermal efficiencies for each month through the year.

thermal equator The belt between 23° N and 10–15° S where the temperature is highest. Its location changes with the seasons; its mean position is about 5° N.

thermal equilibrium The condition of two or more bodies that are at the same temperature. Unless some outside process intervenes, energy will not be exchanged between them.

thermal high An area of high pressure produced by the cooling of air in contact with a cold surface.

thermal inertia A measure of the rate at which a material responds to a change in the amount of energy it absorbs. Substances with a high heat capacity, such as water, respond very slowly. Those with a lower heat capacity, such as rocks and dry sand, respond more rapidly.

thermal-infrared radiation Infrared radiation with a wavelength of about 3–15 μm.

thermal low (heat low) An area of low pressure produced by the warming of air in contact with a warm surface.

thermal pollution The release into the environment of air or water that is markedly hotter than the air or water into which it is discharged.

thermal resistivity A measure of how poorly a material conducts heat.

thermal soaring Circling in the rising air of a thermal to gain height with the minimum expenditure of energy.

thermal steering The movement of an atmospheric disturbance in the direction of the nearest thermal wind.

thermal tide A change in air pressure produced by the heating of the surface by sunshine. It follows the progress of the Sun and therefore progresses around the Earth in a similar fashion to the gravitational tides.

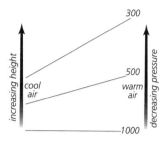

thermal wind

thermal wind A wind, e.g., the jet stream, generated when the air temperature changes by a large amount over a short horizontal distance. Where warm and cool air lie side by side, atmospheric pressure decreases with height more rapidly in the cool than in the warm air, because the cool air is more compressed. Consequently, a surface of constant atmospheric pressure slopes upward from the cool to the warm air and the thickness of each atmospheric layer increases along a gradient from the cool to the warm air. This gradient increases with height, because compression decreases with height more slowly in cool air than in warm air. The geostrophic wind speed is proportional to the gradient, so if the gradient increases with height, so must the wind speed. The thermal wind blows with the cool air to its left in the Northern Hemisphere and to its right in the Southern Hemisphere.

thermistor A thermometer that measures electrical resistance in a material in which resistance varies with the temperature.

thermodynamic diagram A diagram summarizing the factors affecting the temperature and pressure of a parcel of air. A point on the diagram refers to the thermodynamic (energy) state of the air in that location. A simple thermodynamic diagram measures altitude along its vertical axis, temperature along its horizontal axis.

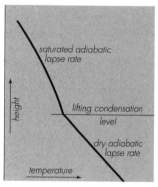

thermodynamic diagram

thermograph An instrument that provides a continuous record of temperature.

thermometer An instrument used to measure the temperature of a substance. Thermometers are used in meteorology to measure the temperature of air and water.

thermosphere The uppermost layer of the atmosphere, lying above the mesopause. Its lower boundary is about 56 miles (90 km) above sea level. It has no precise upper margin.

thermotropic model A model of the atmosphere used in numerical forecasting to forecast the height of one constant-pressure surface, usually at 500 mb, and the height of one temperature, usually the mean temperature between 1,000 mb and 500 mb. From these it is possible to construct a forecast surface chart.

thickness The difference in altitude between two isobaric surfaces.

thickness chart A synoptic chart showing the changing thickness of a particular atmospheric layer.

thickness line (relative isohypse) A line drawn on a map to join places where the thickness of a given atmospheric layer is the same.

thickness pattern (relative hypsography) The pattern made by the thickness lines on a thickness chart.

Thomson effect A water molecule at the surface of a body of liquid with a plane (level) surface is bound more tightly than one in a spherical droplet, and the smaller the droplet, the more easily the molecule can escape.

Thornthwaite climate classification A scheme for classifying climates that uses temperature and precipitation to define the boundaries of climatic types, based on the concept of potential evapotranspiration and a moisture index.

three-cell model A description of the general circulation of the atmosphere that is a good approximation of the way the atmosphere behaves. It comprises a system of vertical cells of three types. Hadley cells operate in the Tropics and polar cells in high latitudes. These are direct cells, driven by convection, and they drive the indirect Ferrel cells lying between them.

three-front model A model of the distribution of air masses over North America that includes three approximately parallel fronts running from northwest to southeast and marking the boundaries between maritime arctic, maritime polar, continental tropical, and maritime tropical air.

threshold limit value (TLV) The greatest concentration of a specified airborne pollutant to which workers may be legally exposed day after day.

threshold velocity The minimum speed at which wind raises dust, soil, or sand particles from the ground.

throughfall The proportion of the total precipitation falling over a forest that reaches the ground.

thunder The sound caused by the discharge of energy during a lightning flash, as air is heated rapidly and expands explosively.

thunderbolt A rock, piece of metal, or dart that storm gods were believed to hurl and that was believed to cause the damage in fact due to lightning. Lightning strikes often melt sand grains, producing irregular masses of glass that were assumed to be thunderbolts.

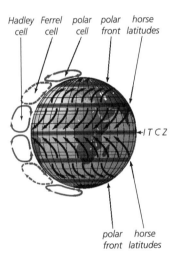

Hadley Ferrel polar polar horse
cell cell cell front latitudes

ITCZ

polar horse
front latitudes

three-cell model

thunderhead A cumulonimbus cloud that extends to the tropopause, where it spreads downwind to produce an anvil-shaped incus.

thunderstorm A violent storm that causes heavy precipitation, strong gusts of wind, and thunder and lightning.

thunderstorm day A day on which a thunderstorm is observed at a weather station.

Tibetan high An anticyclone that develops in summer over the Tibetan Plateau.

tidal wind (1) A wind produced by the atmospheric tides. (2) A wind produced in some tidal inlets by the displacement of air when the tide rises strongly.

tipping-bucket gauge A rain gauge in which rainwater entering a funnel is guided to a second, smaller funnel and from there into one of two buckets mounted on a rocker. When the bucket is full its weight tips it downward, making an electrical contact that is transmitted to a recording pen, which moves on a graph mounted on a rotating drum. That bucket empties and the second bucket is positioned to collect water.

tjaele (frost table) A layer of frozen ground, or ground containing frozen water, at the base of the active layer in a permafrost region.

tongara A hazy southeasterly wind that blows through the Makasar Strait, Indonesia.

Topex-Poseidon A joint U.S.–French satellite mission to measure the height of the surface of the ocean with unprecedented accuracy.

tornadic storm A storm capable of generating tornadoes.

tornado (1) The most violent of all weather phenomena, comprising a rapidly rotating vortex that extends downward from a large storm cloud; it becomes a tornado when it touches the ground. (2) A violent but brief thunderstorm in West Africa.

Tornado Alley An area of the U.S. Great Plains where tornadoes occur more frequently than anywhere else in the world and where there are the most violent tornado outbreaks.

Torrid Zone The climatic region between the tropics of Cancer (23.5° N) and Capricorn (23.5° S).

totable tornado observatory (TOTO) A package of instruments designed to measure wind speed, atmospheric pressure, temperature, and electrical discharges inside a tornado with winds up to 200 mph (322 kmh^{-1}).

total ozone mapping spectrometer (TOMS) A satellite-borne instrument that measures ozone concentrations and provided the first evidence of stratospheric ozone depletion over Antarctica. A second TOMS was launched in 1991 and a third in 1996.

touriello A southerly wind, of the föhn type, that descends from the Pyrenees and blows along the Ariège Valley, in southwestern France.

towering Vertical stretching of a mirage caused by the downward curvature of light due to refraction increasing with altitude.

trade cumulus (trade-wind cumulus) Cumulus cloud that forms over the tropical oceans in air trapped beneath the trade wind inversion.

trade wind inversion An inversion associated with the subsidence of air on the high-latitude side of the Hadley cell circulation.

trade winds The winds that blow toward the equator from the northeast in the Northern Hemisphere and from the southeast in the Southern Hemisphere. They are extremely dependable, especially on the eastern side of the Atlantic, Pacific, and Indian Oceans, blowing at an average speed of about 11 mph (18 kmh^{-1}) in the Northern Hemisphere and 14 mph (22 kmh^{-1}) in the Southern Hemisphere.

tramontana A northerly or northeasterly wind in the northwestern part of the Mediterranean region.

transfrontier pollution The movement of pollutants that are carried in air or water across an international frontier.

translucidus A variety of translucent clouds that cover a large proportion of the sky but through which it is possible to discern the position of the Sun or Moon.

transmissivity A measure of the transparency of the atmosphere to incoming solar radiation.

transparency The capacity of a medium for permitting radiation to pass through it with no significant scattering or absorption.

transparent sky cover The proportion of the sky covered by cloud that does not completely hide whatever may be above it.

transpiration The evaporation of water through the surface pores in plant leaves (stomata) and stems (lenticels).

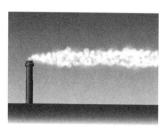

trapping

trapping The pattern made by a chimney plume that widens a little and descends slightly as it moves downwind, but with little dispersion of the gases and particles.

traveling wave A wave that moves through a medium, although the particles composing the medium oscillate about a fixed point.

tree line The climatic limit beyond which temperatures are too low to permit trees to grow.

tropical air Warm air originating over oceans in the subtropical high-pressure belt, over continents at the edge of this belt, or in the interior of continents in summer, when subsiding air produces high surface pressure and tropical air masses.

Tropical Atmosphere Ocean (TAO) A monitoring network that uses an array of moored buoys to gather data from the sea surface and sea-level atmosphere.

tropical cyclone An extensive area of low surface pressure that develops in the Tropics and generates fierce winds and rain to become the biggest and most violent type of atmospheric disturbance experienced on Earth. A tropical cyclone will develop only if the sea-surface temperature is at least 80°F (27°C) over a large area, if there is an area of low pressure no closer to the equator than 5°, and if there is wind shear at a high level to remove the rising air.

Tropical Cyclone Program A program run by the United States National Weather Service that coordinates monitoring and research into tropical cyclones. It focuses mainly on increasing the accuracy of predicting the tracks and intensity of tropical cyclones.

tropical depression An area of low pressure that develops in the Tropics through the convergence of air at low level.

tropical disturbance An incipient tropical storm caused by the convergence of air at a low level and not associated with a frontal system.

Tropical Ocean Global Atmosphere (TOGA) A program that ran from 1985 until 1994 with the aim of improving the understanding and prediction of ENSO events.

tropical storm A tropical depression that has deepened until the winds around it are blowing at speeds of 38–74 mph (61–119 kmh⁻¹).

Tropics The two lines of latitude, 23.5° N and S, at which the Sun is directly overhead at noon on one of the solstices, or the region between them. The northern tropic is the tropic of Cancer and the southern is the tropic of Capricorn.

solstice equinox solstice
sun   sun   sun

23°30' 23°30'

23°30' tropic equator 23°30' tropic

Tropics

tropopause The boundary between the troposphere and stratosphere. Its height averages about 10 miles (16 km) over the equator, 7 miles (11 km) in middle latitudes, and 5 miles (8 km) at each pole.

tropopause break A discontinuity in the tropopause through which tropospheric air and water vapor is able to cross into the stratosphere and stratospheric ozone sometimes enters the troposphere.

tropopause chart A synoptic chart showing the vertical distribution of pressure through the troposphere, isotherms, the height of the tropopause, and tropopause breaks.

tropopause fold A downward depression in the tropopause, bringing stratospheric air into the troposphere, sometimes as far as the surface.

troposphere The layer of the atmosphere from the surface to the tropopause, containing virtually all of the atmospheric water vapor, in which winds and convection thoroughly mix the air.

trough A long, tongue-like protrusion of low pressure into an area of higher pressure.

trowal A Canadian term for a trough of warm air held high above the surface by an occlusion.

tuba A supplementary feature of clouds that may give rise to tornadoes or waterspouts; it consists of a tapering, funnel-shaped projection beneath the cloud base.

turbidity A reduction in the transparency of the atmosphere due to haze or air pollution.

turbopause The poorly defined upper boundary of the turbosphere, at a height of about 60 miles (100 km), above which air molecules move by diffusion and form layers according to their weights.

turbosphere The region of the atmosphere in which the equilibrium of the air is maintained mainly by convection. It extends from the surface to a height of about 60 miles (100 km).

turbulent flow (turbulence) The movement of a fluid in which elements of the fluid follow streamlines that cross each other, so the flow passing any particular point changes speed and direction in an irregular and unpredictable fashion.

twister A popular name for a tornado.

typhoon A tropical cyclone that forms over the Pacific.

ubac Sloping ground that faces away from the direction of the equator and therefore is shaded.

turbulent flow

ultraviolet index (UVI) A guide to the intensity of ultraviolet radiation, reported as an index value related to the duration of exposure that will cause sunburn in the most susceptible people (those with pale skin).

ultraviolet radiation (UV) Electromagnetic radiation at wavelengths between about 4 nanometers (nm) and 400 nm.

Umkehr effect A reversal in the relative intensities of light at two wavelengths from directly overhead as the Sun's zenith angle varies. Ozone absorbs one wavelength more strongly than the other, so a series of measurements can be used to determine the vertical distribution of ozone.

uncinus (unc) A species of cirrus clouds consisting of long filaments that end in hooks, in the shape of commas, or in tufts, but not with a protruding upper part.

undercast A complete cloud cover (8 oktas or 10/10) as it appears to, and is reported by, the pilot of an aircraft flying above the cloud.

undulatus A variety of clouds in which sheets, layers, or patches of cloud undulate like waves.

unrestricted visibility Horizontal visibility that is not obstructed or obscured for at least 7 miles (11 km).

unsettled An adjective describing weather conditions that are fine, but may change at any time in the near future with the development of cloud and possibly precipitation.

upper air The upper part of the troposphere.

upper-air chart (upper-level chart) A constant-pressure chart depicting the condition of the atmosphere at a pressure level in the upper troposphere, usually at 925, 850, 400, 300, 250, 200, 150, and 100 mb.

upper atmosphere The atmosphere above the tropopause.

upper-atmosphere dynamics The movement of air in the upper atmosphere, especially at heights above 300 miles (500 km), where the motion consists predominantly of atmospheric tides, atmospheric waves, turbulent flow, and sound waves.

Upper Atmosphere Research Satellite (*UARS*) The NASA satellite carrying instruments to measure the temperature, chemical composition, and winds in and above the stratosphere.

upper front A weather front that is present in the upper air, but does not extend to the surface.

upper mixing layer A region in the upper mesosphere, at about 30–50 miles (50–80 km), where temperature decreases rapidly with height and considerable turbulence mixes the atmospheric constituents.

uprush (vertical jet) The very strong upcurrent found in a rapidly developing cumulus cloud.

upwind In the direction from which the wind is blowing.

upwind effect The increase in precipitation over the land or sea on the upwind side of a range of hills or mountains.

urban boundary layer The layer of air over a city extending from the top of the urban canopy layer to the uppermost limit of the region in which the climatic properties of the air are modified by the surface below.

urban canopy layer The layer of air lying below the level of the rooftops in a city. The climate in this layer is strongly modified by the many microclimates produced by the streets and buildings.

urban canyon A city street lined on both sides by tall buildings, so it resembles a canyon. Wind tends to be funneled along the street, making it windier, especially if it is aligned with the prevailing wind.

urban climate The climate of a large city. This is generally warmer, wetter, dustier, and less windy than the climate in adjacent rural areas.

urban dome The approximately domed shape of the warm air beneath the inversion associated with an urban heat island. This is most pronounced on calm nights when the sky is clear and when the heat island is most strongly developed.

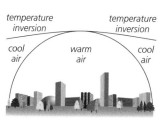

urban dome

vacillation The irregular change in the Rossby waves surrounding each hemisphere as their amplitude oscillates between a maximum and a minimum.

valley breeze A warm, anabatic wind that blows during the day in some mountain areas. As the ground warms, air in contact with it expands and becomes less dense. Cooler, denser air subsides from a higher level, pushing the warm air up the mountainsides.

valley fog Radiation fog that forms at night in the bottom of valleys. As the hillsides cool by radiating the warmth they absorbed during the day, chilled air sinks to the valley floor.

valley breeze

vapor pressure The partial pressure exerted on a surface by water vapor present in the air.

vardarac (vardar) A cold, northwesterly wind resembling the mistral that blows in the fall along the valley of the Vardar River, Macedonia.

variable ceiling A ceiling that changes in height rapidly and repeatedly while its height is being measured.

variable gas An atmospheric gas, the amount of which varies as a proportion of the whole from place to place or time to time. Water vapor, carbon dioxide, and ozone are the most important variable gases.

variable visibility The condition in which visibility increases and decreases rapidly while it is being measured.

vaudaire (vauderon) A strong, southerly föhn wind that blows across Lake Geneva, Switzerland.

vector quantity A physical amount that acts in a direction so that its description must include both the magnitude of the amount and its direction of action.

veering A change in the wind direction in a clockwise direction, for example, from the southwest to the northwest.

vegetation index mapping The use of satellite images to identify areas experiencing deforestation, drought, or desert encroachment.

veil of cloud A very thin layer of cloud. Objects can be seen through it.

velocity A vector quantity stating the speed at which a body is traveling in a specified direction.

velum An accessory feature of clouds, comprising a layer of cloud extending horizontally for a considerable distance above other clouds and sometimes connecting the tops of cumulus clouds.

vendaval A strong, sometimes gale-force, southwesterly wind along the coast of Spain.

vent du Midi A southerly wind that blows through the central region of the Massif Central and Cevennes in the Midi region of southern France.

ventifact A desert pebble that has clearly defined faces due to erosion by wind-blown sand. Provided the pebble is not moved, the positions of the faces indicate the direction of the prevailing wind.

ventilation The removal of pollutants from the air by the action of the wind, which introduces clean air.

Venturi effect The acceleration of the wind when the flow of air is constricted, for example, by tall buildings or hills.

veranillo Along the western coasts of tropical Central and South America, a dry, hot season ("little verano"), lasting a few weeks, that interrupts the long wet season.

verano The name given to the long dry season, from November until April, in parts of tropical Central and South America.

vertebratus A variety of cirrus clouds in which the pattern of the cloud elements is reminiscent of a fish skeleton, with the long ribs clearly displayed.

vertical anemometer An instrument that measures the vertical component of wind speed.

vertical differential chart A diagram showing values for a particular atmospheric feature at two different heights.

virazon A southwesterly sea breeze that blows onto the western coasts of Chile and Peru.

virga (fallstreaks) A wispy veil beneath the base of a cloud that does not reach to the ground. It is precipitation falling into relatively dry air, where it evaporates.

Viroqua A city in Wisconsin that was struck by one of the most severe tornadoes ever recorded, on the afternoon of June 28, 1865. On the basis of the damage it caused, the tornado has been judged F4 on the Fujita tornado intensity scale.

virtual temperature The temperature dry air would have if it were at the same density and pressure as moist air. This is a little higher than its actual temperature.

viscosity The resistance a fluid presents to shear forces, and therefore to flow.

visibility The transparency of air to visible light, measured as the distance from which an observer is able to distinguish an object with the naked eye.

visible radiation Shortwave electromagnetic radiation at wavelengths of 0.4–0.7 μm, to which the human eye is sensitive.

voluntary observing ship (VOS) A merchant ship equipped to act as a weather station.

Volz photometer An instrument for measuring the intensity of direct sunlight. The measurements are not very precise, but readings can be taken by relatively unskilled workers and the sky does not need to be completely cloudless.

von Kármán constant A value used in calculations of wind profiles; its value is approximately 0.4. (It is a constant, and therefore a number, rather than a measurement in units.)

vorticity The tendency of a mass of fluid moving relative to the surface of the Earth to turn about a vertical axis.

vortex A spiraling movement in a fluid, affecting only a local area. The fluid at the center of a vortex is usually stationary or slow-moving, but it may be surrounded by a gas or liquid that is moving very rapidly.

wake A region of turbulence downwind of a surface obstruction or behind a body moving through a fluid.

wake low A small area of low air pressure to the rear of a fully developed thunderstorm.

Walker circulation A slight but continuous latitudinal movement of air as a series of cells (Walker cells) between the equator and about latitude 30° in both hemispheres. Air rises over the western Pacific and eastern Indian Oceans. At high level each rising current separates into two streams flowing east and west. The high-level streams from neighboring cells converge and subside over the eastern Pacific and the western Indian Oceans. The Walker circulation produces a wet climate in Indonesia and a dry climate in western South America. Every few years the pattern changes and the Walker circulation over the Pacific weakens or reverses. This change is associated with the southern oscillation.

wall cloud The extension that appears below the base of a cumulonimbus cloud when a mesocyclone inside the cloud is expanding downward,

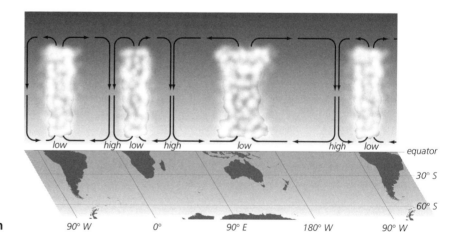

Walker circulation

possibly to become a tornado. The wall cloud marks the region where warm, moist air is being drawn into the main updraft and moisture is condensing. It rotates cyclonically, due to convergence, and releases no precipitation.

warm cloud A cloud in which the temperature is above freezing throughout.

warm front An advancing front with warm air behind it.

warm high (warm anticyclone, warm-core anticyclone, warm-core high) An anticyclone that is warmer at its center than near the edges.

warm low (warm-core cyclone, warm cyclone, warm-core low) A cyclone that is warmer at its center than near the edges.

warm rain Rain that falls from a warm cloud.

warm seclusion A pool of warm air that lies above the center of a cyclone.

warm sector The region containing the wedge of warm air inside the frontal wave of a frontal system, between the warm and cold fronts.

warm water sphere The part of the oceans where the temperature is higher than 46°F (8°C).

warm wave A sudden rise in temperature in middle latitudes, usually occurring in summer and often heralding wet weather.

Wasatch wind A strong easterly wind that blows across the plains of Utah from the mouths of canyons in the Wasatch Mountains.

washout The removal of solid particles from the air by collision with falling raindrops.

water balance (moisture balance) The difference between the amount of water that reaches an area as precipitation and the amount lost by evapotranspiration and runoff. It is calculated for a specified area and for a specified time (usually one year) by: $p = E + f + \Delta r$, where p is precipitation, E is evapotranspiration, f is filtration (the amount of water absorbed and retained by the soil), and Δr is net runoff.

water cloud A cloud composed entirely of water droplets.

water deficit The difference between the amount of water needed to sustain healthy plant growth (usually of crop plants) and the amount supplied by precipitation and available to plants, where this is smaller than the amount required.

water devil A phenomenon resembling an aquatic dust devil, but smaller, less violent, and of shorter duration.

water equivalent The depth of a fall of snow after this has been converted into an equivalent depth of rainfall.

water sky The dark appearance of a cloudy sky over water, when the Sun is high and little light is reflected to illuminate the cloud base.

waterspout A column of spiraling air over water that resembles a tornado and is larger than a water devil.

water surplus The difference between the amount of water needed to sustain healthy plant growth (usually of crop plants) and the amount supplied by precipitation and available to plants, where this is greater than the amount required.

wave depression (wave cyclone) A depression that forms around the point where a cold front and warm front meet and warm air is beginning to rise over, or be undercut by, the cold air.

wave equation A partial differential equation describing the velocity v and vertical displacement produced by a wave (ψ) as a function of space and time (t), where space is described by three coordinates $x, y,$ and z. The equation can be written as: $v^2\psi = \delta^2\ \psi/\delta x^2 + \delta^2\ \psi/\delta y^2 + \delta^2\ \psi/\delta z^2 = (1/v^2)\delta^2\ \psi/\delta t^2$.

wave front A line joining all the points that are at the same phase along the path of an advancing wave.

wave hole A hole in a layer of cloud on the lee side of a mountain, caused by a stream of air descending from the peak, warming adiabatically and causing cloud droplets to evaporate.

weakening A decrease in the pressure gradient, causing a reduction in wind speed.

weak front A warm front that is overriding a mass of cold air, but behind which the relative humidity is low.

weather The state of the atmosphere in a particular place at a particular time or over a fairly brief period, with special emphasis on short-term changes.

weather glass A barometer, in particular a household one mounted on the wall.

weather house A hygrometer in the form of a household ornament comprising a house with two doors and a figure behind each. The figures are made to move by the contraction or stretching of one or two lengths of hair in response to the humidity. When one of the figures emerges it is supposed to mean rain is imminent and when the other figure emerges the weather is supposed to remain fine.

weathering All the natural processes by which solid rock is broken into ever smaller fragments. Not all weathering processes are associated with weather.

weather lore The accumulated traditions and observations with which our ancestors attempted to interpret weather signs and forecast the weather.

weather map A map showing the distribution of pressure, wind, and precipitation over an area of the Earth's surface at a particular time.

weather minimum The poorest weather conditions in which aircraft are permitted to fly under visual (VFR) or instrument (IFR) flight rules.

weather observation A record of weather conditions based on measurements made in a standard fashion and written down according to a strict formula.

weather radar Radar used to study processes inside clouds, especially the density of water droplets, where the water is most concentrated, and the level at which rising water droplets freeze and falling ice melts.

Weather Radio A network of more than 480 stations that broadcast continuous weather information over the whole of the United States, U.S. coastal waters, Puerto Rico, the U.S. Virgin Islands, and the U.S. Pacific Territories.

weather satellite A satellite in Earth orbit carrying instruments that produce images from which meteorological and climatological information can be obtained.

weather ship A weather station mounted on a ship that is anchored permanently in one location (except when it needs to return to port for repairs or maintenance).

weather shore The shore from which the wind is blowing, as seen by a ship at sea.

weather side The side of a ship facing into the wind or weather.

weather station A place equipped with the instruments needed to make standardized measurements and observations of weather conditions, and the technical and communications facilities to transmit weather reports to a central point.

wedge A ridge of high pressure in which the isobars make a V-shaped point, like a wedge.

Wefax An abbreviation for "weather facsimile," which is a system for transmitting by radio such material as graphic reproductions of weather maps, summaries of temperatures, and cloud analyses. Most Wefax transmissions are from GOES craft.

weir effect The movement of air that is carried inland by sea breezes in western South America from northern Peru to central Colombia. Air rises up the main part of the range and spills over the top, like water over a weir, pushed by the air behind it, and descends into the valleys as a katabatic wind.

westerly type Weather generated and transported by the prevailing westerlies and characteristic of middle latitudes.

westerly wave A wave-like disturbance embedded within the prevailing westerlies of middle latitudes.

western intensification The strengthening of ocean currents as they flow along the western coasts of the continents.

wet-bulb depression The difference between the temperature registered by a dry-bulb thermometer and that registered by a wet-bulb thermometer adjacent to it.

wet-bulb potential temperature The wet-bulb temperature a parcel of saturated air would have if it were taken adiabatically to the 1,000 mb level (i.e., to sea level).

wet-bulb temperature The temperature registered by a wet-bulb thermometer.

wet-bulb thermometer A thermometer with a layer of wetted cloth around its bulb and extending into a reservoir of distilled water, so it acts as a wick. Water is drawn into the wick and evaporates from it, taking latent heat of vaporization from the thermometer bulb, thus depressing the temperature registered by the thermometer.

wet spell In Britain, a period of at least 15 days during which at least 0.04 inch (1 mm) of rain has fallen every day.

wettability The property of a surface that has an affinity for water, because its molecules attract water molecules.

whaleback cloud A type of lenticular cloud with a smooth, humped top that is sometimes seen in high latitudes. It forms in strong winds over islands and steep coastal cliffs.

whirling psychrometer (sling psychrometer) A psychrometer in which the flow of air across the bulbs of the dry-bulb and wet-bulb thermometers is produced manually by whirling the instrument through the air.

whirlwind A large dust devil that appears suddenly when the air is calm. Most whirlwinds are short-lived, but some last several hours.

whirling psychrometer

whirly A very small, violent storm in Antarctica, most common around the time of the equinoxes.

white dew Dew that freezes after it has formed.

white horizontal arc An arc very occasionally seen when sunshine falls onto a cloud consisting of ice crystals with both vertical and horizontal faces. The arc passes through the shadow of the observer, which is cast onto the cloud.

white noise Random data that obscure the signal from a measuring device.

whiteout The condition in which the ground, air, and sky are all a uniform white and no landscape features are visible.

white squall A sudden squall that occurs in tropical or subtropical seas without the prior appearance of a squall cloud. A line of white water caused by the wind is the only indication of its approach.

whole gale A wind of 55–63 mph (88–101 kmh^{-1}). It is force 10 on the Beaufort wind scale.

Wien's law A law describing the relationship between the temperature of a black body and the wavelength (only at short wavelengths) of its maximum emission of radiation. The law can be stated as: $\lambda_{max} = C/T$, where λ_{max} is the wavelength of maximum emission, C is Wien's constant, and T is the temperature in kelvins. Wien's constant is $2,897 \times 10^{-6}$m, so the law becomes: $\lambda_{max} = (2897/T) \times 10^{-6}$m.

wild snow Freshly fallen snow that is very light and unstable.

williwaw (1) A violent squall in the Strait of Magellan. (2) A strong squall wind that blows as cold air down deep valleys in Alaska.

willy-willy (1) An Australian name for a dust devil or whirlwind. Willy-willys typically occur on hot days in dusty areas in the outback. (2) The name that was formerly given to a tropical cyclone that strikes northwestern Australia. The name fell into disuse early in the 20th century.

wind chill The extra feeling of cold people experience when exposed to the wind, due to the removal of warm air from the body surface. The effect can be measured and its magnitude is usually reported in degrees Fahrenheit (or Celsius).

wind field A pattern of winds associated with a particular distribution of pressure.

wind imaging interferometer (WINDII) An interferometer carried on the *Upper Atmosphere Research Satellite* that calculates temperatures and winds in the thermosphere.

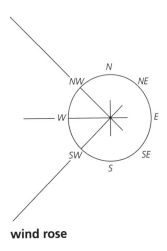

wind rose

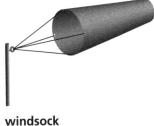

windsock

wind profile A diagram showing the change in wind characteristics with height and horizontal distance. This can be used to show the effect on the wind of an obstruction. The wind speed at any height in the profile is given by $u = (u^*/k) \ln (z/z_0)$, where u is the wind speed, u^* is the friction velocity, k is the von Kármán constant, z is the height above the surface, and z_0 is the roughness length (ln means "natural logarithm").

wind profiler A ground-based instrument that measures the wind speed at different heights and at frequent intervals.

wind ripple (snow ripple) A wave-like pattern at right angles to the wind direction on the surface of snow, produced by the wind.

wind rose A diagram showing the frequency with which the wind at a particular place blows from each direction.

windrow Loose material that has accumulated naturally to form a line. If it occurs inland or on the surface of open water, the material has been arranged by the wind and the orientation of the line indicates the wind direction. If it occurs on a beach, it has been formed by the action of the tides.

wind scoop A shallow depression in the surface of snow or loose sand at the base of walls and close to trees and other obstructions; it is caused by eddies.

wind shadow thermal A strong thermal that develops in warm, sunny weather wherever turbulence greatly reduces the wind speed close to the surface.

wind shear A change of wind velocity with vertical or horizontal distance.

wind-shift line A boundary marking a large and abrupt change in the wind direction.

windsock A device for indicating the direction of the wind, comprising a tapering fabric cylinder open at both ends, supported on a circular frame at the larger end and attached by cords to the top of a tall pole.

windstorm A storm in which the most significant characteristic is a very strong wind.

wind stress The force per unit area exerted on the land or water surface by the movement of air.

wind vane The most commonly used device for measuring the direction of the wind, consisting of four fixed arms pointing to the cardinal points of the compass and above them the vane, with a flat surface on one side and a pointer on the other, mounted so it can move freely around a vertical axis. The wind pushes the flat surface to the far side of its axis. Consequently, the pointer indicates the direction from which the wind is blowing. That is why we name winds by the direction from which they blow and not by the direction in which they are blowing.

wind velocity The speed and direction of the wind when these are reported together.

windward The side facing the direction from which the wind is blowing.

winter ice A layer of ice less than one year old and more than 8 inches (20 cm) thick that forms a complete covering on a large area of the sea surface.

World Climate Program (WCP) A program established by the World Meteorological Organization in 1979 to collect and store climate data.

World Meteorological Organization (WMO) The specialized United Nations agency that exists to promote the establishment of a worldwide system for gathering and reporting meteorological data, the standardization of the methods used, the development of national meteorological services in less-industrialized countries, and the application of meteorological information and understanding to other fields.

xenon A rare gas, heavier than the other atmospheric gases, that accounts for about 0.0000086% of the atmosphere (0.1 part per million by volume). Atomic number 54; atomic weight 131.30; density (at sea-level pressure and 32°F, 0°C) 0.059 ounces per cubic foot (5.887 grams per liter). It melts at −169.6°F (−111.9°C) and boils at −160.6°F (−107.1°C).

yamase A cool, onshore, easterly wind in the Senriku district of Japan.

Younger Dryas (Loch Lomond stadial) A cold, dry period, affecting the whole of northern Europe but not North America, that lasted from about 11,000 to 10,000 years ago. Soils of that age contain abundant pollen from mountain avens *(Dryas octopetala),* an arctic-alpine plant.

young ice A layer of ice less than one year old and more than about 2 inches (5 cm) thick that forms a complete covering on a large area of the sea surface.

Young's modulus The ratio of the stress applied to a solid and the change in its length.

zenithal rain Rain that falls in the Tropics or subtropics every year, or every other year, during the season when the Sun is most nearly overhead (at its zenith).

zenith angle The height of the Sun above the horizon, measured as the angle between a line linking the observer to the Sun and the vertical (the zenith). The zenith angle *(Z)* is calculated from cos Z = sin θ sin δ + cos θ cos δ cos h, where θ is the latitude, δ is the solar declination, and h is the hour angle (15 (12 – t), where t is the local time using a 24-hour clock).

zenith transmissivity A measure of the fraction of solar radiation reaching the top of the atmosphere when the Sun is directly overhead (at its zenith) that penetrates to the surface.

zephyr Any very gentle breeze, but especially one that blows from the west.

zobaa A very tall whirlwind of sand, resembling a pillar, sometimes seen in the Egyptian desert. It moves very rapidly.

zodiacal light A band of light that appears in the sky at night, in the east shortly before dawn or in the west shortly after sunset.

zonal circulation The west-to-east movement of air in the troposphere, measured over a specified period or with respect to a particular longitude.

zonal flow A movement of air, generally in a west-to-east direction and approximately parallel to the lines of latitude.

zonal index The strength of the westerly winds between latitudes 33° and 55° in both hemispheres, expressed either as the horizontal pressure gradient or as the corresponding geostrophic wind. A high zonal index indicates strong westerly winds and a continuous and almost straight jet stream. A low index indicates weak westerlies and a cellular air-flow pattern.

zonal kinetic energy The kinetic energy of the mean zonal wind.

zonal wind-speed profile A diagram in which the speed of the zonal flow is plotted against latitude.

zonda A warm, dry föhn wind that occurs in winter in the lee of the Andes in Argentina.

zone of maximum precipitation The elevation on a mountainside where the precipitation is greatest.

SECTION TWO
BIOGRAPHIES

Abbe, Cleveland (1838–1916) An American meteorologist, who was the first person to issue regular daily weather bulletins and forecasts and who is sometimes called the "father of the Weather Bureau," Cleveland was born on December 3, 1838, in New York City. He was also influential in establishing standardized time zones across the United States. After graduating in astronomy, in 1866 he was appointed director of the Cincinnati Observatory, from where he began issuing a daily *Weather Bulletin* and weather forecasts in 1869. This led to the establishment of the National Weather Bureau in 1870 (renamed the United States Weather Bureau in 1891). Abbe became meteorologist in charge, retaining the position for the rest of his life, and combining it with research and teaching at Johns Hopkins University. In 1879 he wrote a report recommending the standardization of time zones. This was adopted in 1883 in the United States and later extended to the whole world. Cleveland Abbe died on October 28, 1916, in Chevy Chase, Maryland, where he and his wife are buried.

Agassiz, Louis (1807–73) This Swiss–American naturalist, who established that at one time much of the Northern Hemisphere had been covered by ice sheets and glaciers during a "Great Ice Age," was born on May 28, 1807, at Motier, near Friborg, Switzerland. Agassiz began his career as an authority on the classification of fishes, including fossil fishes, as well as fossil echinoderms and mollusks. In 1836 he turned to the study of erratics (boulders found far from the rock formations from which they have been detached). Some scientists suspected that glaciers might have transported them. Agassiz proved that the glaciers flowed, thus confirming that Swiss glaciers had once extended deep into France, carrying the erratics before them. In 1846 he visited the United States to deliver a series of lectures and remained in the country. In 1848 he was appointed professor of zoology at Harvard University. He became an American citizen and remained in the United States for the rest of his life, most of the time at Harvard, where in 1858 he developed the Museum of Comparative Zoology to assist research and teaching, building it around his own collection. He was its director from 1859 until his death. In addition to his research, Louis Agassiz was one of the finest teachers of science America has ever known. He was devoted to his students and treated them as collaborators. He died at Cambridge, Massachusetts, on December 12, 1873. In 1915 Agassiz was elected to the Hall of Fame for Great Americans.

Aitken, John (1839–1919) A Scottish physicist, who was born and died at Falkirk, in Stirlingshire, Aitken discovered that the air contains large numbers of very small particles, now known as Aitken nuclei, and invented the Aitken nuclei counter, an instrument for detecting them. He also discovered the part Aitken nuclei play, as cloud condensation nuclei, in the formation of clouds. Poor health made it impossible for him to hold any official position. Instead, he worked from a laboratory he made at his home, where he constructed his own apparatus and conducted experiments. He described many of his findings in papers published in the journals of the Royal Society of Edinburgh, of which he was a member.

Amontons, Guillaume (1663–1705) A French physicist, and one of the most ingenious inventors of his age, Amontons was born in Paris on August 31, 1663. In 1687 he invented a new type of hygroscopic hygrometer, based on the expansion and contraction of a substance as it absorbs and loses atmospheric moisture. In 1695 he invented a barometer that did not require a reservoir of mercury, which meant it could be used at sea. The same year he improved on the air thermoscope that had been invented by Galileo in 1593. He used his thermometer to show that water always boils at the same temperature and that the volume occupied by a gas always changes by the same amount for a particular change in temperature. He was also able to visualize a temperature at which gases contracted to a volume beyond which they could contract no further. This was the concept of absolute zero. So far as is known he did not attend a university, and from his teens he was profoundly deaf. He earned his living as a government employee, working on a range of public works. In 1690 he became a member of the Académie des Sciences. He published a number of papers and one book, which appeared in 1695. Amontons died in Paris on October 11, 1705.

Aristotle (384–322 BCE) A Greek philosopher born in 384 BCE at Stagirus, a Greek colony on the coast of Macedonia, Aristotle wrote about logic, ethics, politics, biology, physics, astronomy, and many other topics. His writings are contained in 47 surviving works. In one of these, *Meteorologica,* Aristotle set out his own explanations for the weather and gave us our word "meteorology." He proposed theories to explain the formation of clouds, rain, hail, wind, thunder, lightning, and storms. Aristotle had no instruments to measure atmospheric conditions, nor any way to compile a picture of weather conditions over a large area. Consequently, he was not able to develop an accurate understanding of atmospheric processes. The importance of his contribution to scientific thinking arises from his insistence on basing theories on observed facts rather than tradition or unsupported opinion. Following the death of his father, in about 367 BCE, when he was 17, his guardian Proxenus sent him to the Academy in Athens, led by the philosopher Plato (428 or 427–348 or 347 BCE). He remained at the Academy for 20 years, first as a pupil and then as a teacher. Philip, the king of Macedonia, appointed Aristotle to supervise the education of his son, a 13-year-old boy who would later become known as Alexander the Great. Later in his life Aristotle was very wealthy, possibly from the money he was paid for teaching Alexander. Aristotle returned to Athens in about 335 BCE and for the next 12 years he taught at the Lyceum, one of the three most famous schools in the city. There he began to assemble a library of books and maps and a museum of natural history. He also established a zoo with animals captured during Alexander's campaigns in Asia. In 323 BCE Aristotle moved to Chalcis (now called Khalkis) on the Greek island of Euboea, north of Athens, where he died the following year. He was 62 years old.

Arrhenius, Svante (1859–1927) A Swedish physical chemist, Arrhenius was born on February 19, 1859, on the estate of Vik, near Uppsala, Sweden, owned by the University of Uppsala. In 1896 he published a paper, "On the Influence of Carbonic Acid in the Air upon the Temperature of the Ground" (*Philosophical Magazine,* vol. 41, pages 237–271). In it, he calculated the effect carbon dioxide would have if the atmospheric concentration of it were altered. He worked out the resulting change in mean temperature for 13 belts of latitude, each of 10 degrees from 70° N to 60° S, for the four seasons of the year and the mean for the year. He calculated that a doubling of atmospheric carbon dioxide would increase the mean annual temperature by 4.95°C at the equator and by 6.05°C at 60° N. For his earlier work on electrolytes, Arrhenius received the 1903 Nobel Prize for chemistry. He also received many other awards and honorary degrees. In 1891 Arrhenius accepted a lectureship at the Stockholms Högskola, where in 1895 he became professor of physics. From 1897 until 1905 Arrhenius was also rector. In 1905 the Swedish Academy of Sciences established a Nobel Institute for Physical Chemistry and Arrhenius was appointed director. He remained in this position until shortly before his death. He died in Stockholm on October 2, 1927, and is buried in Uppsala.

Beaufort, Francis (1774–1857) The British scientist who devised the scale of wind force that bears his name was born on May 7, 1774, at Navan, County Meath, Ireland. In 1806 Commander Beaufort, as he was then, drew up his "Wind Force Scale and Weather Notation." Its aim was to provide guidance for sailors, telling them how much sail they should set on a full-rigged warship according to the wind, but in its first version it included no information about the actual speed of the wind. The scale was introduced throughout the Royal Navy in 1838. In 1874, modified to include details of the state of the sea and the visible effects of the wind on land, the scale was adopted by the International Meteorological Committee for use in international meteorological telegraphy. Beaufort joined the Royal Navy as a cabin boy at the age of 16. By the time he was given his first command, in 1805, he had also become a hydrographer. He conducted surveys in South American and Turkish waters and while at sea he came to recognize the importance of weather conditions and the value of recording them. In 1829 he was made the official hydrographer for the Admiralty. Beaufort was knighted for his service in 1848 and, by the time he retired in 1855, he had reached the rank of rear admiral. He died in London on December 17, 1857.

Beckmann, Ernst Otto (1853–1923) The German chemist who invented the Beckmann thermometer was born on July 4, 1853, in Solingen. He worked as a pharmacist's assistant before studying chemistry and pharmacy at Leipzig University. He taught at the Technische Hochschule in Brunswick, and was a professor of physical chemistry at Leipzig, Giessen, and Erlangen Universities, before being appointed director of the Laboratory of Applied Chemistry at Leipzig. In 1912 he became the first director of the new Kaiser Wilhelm Institute for Physical Chemistry and Electrochemistry in Berlin-Dahlem. He died in Berlin on July 13, 1923.

Bentley, Wilson Alwyn (1865–1931) An American farmer who became famous for his photographs of snowflakes, Bentley was born on February 9, 1865, at the family farm in Jericho, Vermont. Using a microscope belonging to his mother, Bentley studied and drew snowflakes, dewdrops, frost, and hailstones. Finding this unsatisfactory, he acquired a bellows camera and microscope objective that allowed him to photograph them. All of his photomicrographs were taken with this original camera. Over the next 50 years he accumulated an archive of more than 5,000 images and became widely known as "the Snowflake Man." He also devised a method, which is still used, for measuring the size of raindrops by exposing a dish containing a layer of sifted flour, about one inch (25 mm) thick. In 1898 Bentley published his first magazine article, in *Popular Scientific Monthly.* After that, he wrote many popular articles and scientific papers. *Snow Crystals,* his only book, was published in 1931 with William J. Humphreys, the chief physicist at the U.S. Weather Bureau, who persuaded him to do it. In 1924 he received the first research grant ever to be awarded by the American Meteorological Society. He died at the farm from pneumonia on December 23, 1931.

Bergeron, Tor Harold Percival (1891–1977) A Swedish meteorologist who, in 1935, proposed a mechanism for the formation of raindrops in cold clouds, in which ice crystals grow and form snowflakes by gathering water at the expense of supercooled water droplets. As the snowflakes fall into warmer air they melt and reach the surface as water droplets. This was later confirmed experimentally by the German meteorologist Walter Findeisen, and it is now known as the Bergeron-Findeisen (or Wegener-Bergeron-Findeisen) mechanism. Bergeron was born on August 15, 1891, at Godstone, near London, England, and educated at the universities of Stockholm, Sweden, and Leipzig, Germany. Bergeron spent the years from 1918 until 1921 as a student and collaborator of Vilhelm Bjerknes at the Bergen Geophysical Institute and after qualifying he joined the staff. He obtained his Ph.D. from the University of Oslo, Norway, in 1928. Bergeron taught at the University of Stockholm from 1935 until 1945 and in 1946 he moved to the University of Uppsala, in Sweden. He was professor of meteorology at Uppsala from 1947 until 1961. He died at Stockholm on June 13, 1977.

Bernoulli, Daniel (1700–82) This Swiss natural philosopher, who showed the link between the pressure of a fluid and its velocity, devised an equation of state, and related atmospheric pressure to altitude, was born at Groningen, the Netherlands, on February 9, 1700, into a family of eminent mathematicians. He enrolled at the University of Basel when he was 13 and specialized in philosophy and logic, passing his baccalaureate examination at 15. When he was 16 he obtained his master's degree. He went on to study medicine at Basel and qualified as a doctor in 1721, although mathematics interested him more. In 1725 he accepted a professorship in mathematics at the Russian Academy in St. Petersburg. He returned to Basel in 1733 as professor of anatomy and botany and in 1750 he became professor of natural philosophy, retaining this post until he retired in 1777. Bernoulli published his most important work, *Hydrodynamica* (Hydrodynamics), in 1738. In it he discussed the theoretical and practical aspects of pressure, velocity, and equilibrium in fluids. The relationship between pressure and velocity is now known as the Bernoulli principle, and its consequences as the Bernoulli effect. Daniel Bernoulli received many honors and was elected to most of the scientific academies of Europe. He died in Basel on March 17, 1782.

Bjerknes, Jacob Aall Bonnevie (1897–1975) The Norwegian-born American meteorologist who discovered that depressions originate as waves on fronts was born in Stockholm on November 2, 1897. During World War I, he helped his father, Vilhelm, organize the network of weather stations that supplied them with the data they used to develop their theories of air masses and polar fronts. Jacob moved to the United States in 1939 and in 1940 became professor of meteorology at the University of California–Los Angeles. He became a U.S. citizen in 1946. Bjerknes also conducted extensive studies of the upper atmosphere and jet stream. He was awarded the gold medal of the World Meteorological Organization in 1959. Jacob Bjerknes died in Los Angeles on July 7, 1975.

Bjerknes, Vilhelm Frimann Koren (1862–1951) This Norwegian physicist and meteorologist, who was one of the founders of modern meteorology and scientific weather forecasting, was born in Oslo on March 14, 1862. He spent two years as a lecturer at the School of Engineering in Stockholm and in 1895 was appointed professor of applied mechanics and mathematical physics at the University of Stockholm. In 1897 Bjerknes proposed a system for forecasting weather scientifically, and in 1904 he published a scientific paper outlining a method of numerical forecasting. Bjerknes returned to Norway in 1907 as a professor at Kristiania University and in 1912 he was appointed professor of geophysics at the University of Leipzig. While there he founded the Leipzig Geophysical Institute. In 1917 Bjerknes returned to Norway to found the Bergen Geophysical Institute as part of the Bergen museum. It is now part of the University of Bergen. He joined the staff of the University of Oslo in 1926 and remained there until his retirement in 1932. While he was at the Bergen Institute, Bjerknes and his colleagues established a network of weather stations throughout Norway and used their observations and measurements to produce general pictures of weather conditions at particular times over a wide area. Studying these led to the discovery of air masses and fronts and to the mechanism by which cyclones develop over the North Atlantic. Bjerknes described this work in 1921 in a book that became a classic: *On the Dynamics of the Circular Vortex with Applications to the Atmosphere and to Atmospheric Vortex and Wave Motion.* This formed the basis for the modern theory and practice of meteorology. Bjerknes died in Oslo on April 9, 1951.

Black, Joseph (1728–1799) The Scottish chemist and physician who discovered latent heat and that carbon dioxide is a normal constituent of air was born in Bordeaux, France, on April 16, 1728. In 1740 he was sent to Belfast to be educated and from there he went to Glasgow University, where he studied medicine and natural sciences. He moved to Edinburgh University in 1751 to complete his medical studies, and in 1754 he submitted a thesis for his doctor's degree in which he described the release of a gas he called "fixed air" (carbon dioxide) when magnesium carbonate is heated and the way "fixed air" combines with calcium oxide. In 1756 Black returned to Glasgow as lecturer in chemistry and he was also appointed professor of anatomy at Edinburgh University, although he exchanged the post for that of professor of medicine. Around 1760 Black found, by careful measurement, that when ice is warmed it melts slowly, but its temperature does not change. He concluded that it was absorbing a quantity of heat that must have combined with the particles of ice and become latent in its substance. He called this latent heat and at the end of 1761 he verified its existence experimentally. In 1766 he became professor of chemistry at Edinburgh University. Joseph Black died in Edinburgh on November 10, 1799.

Bouguer, Pierre (1698–1758) A French physicist and mathematician who studied the intensity of sunlight and the effect on light of its passage through the atmosphere. Born at Croisic, Britanny, on February 16, 1698, Bouguer was the son of a hydrographer, who taught him about the geography of fresh and salt water. Pierre was a child prodigy and began teaching hydrography at Le Havre when he was only 15. He also compiled tables of atmospheric refraction and invented the heliometer, an instrument used to measure the diameter of the Sun and the angular distance between stars. Bouguer became professor of hydrography first at Croisic and later, in 1730, at Le Havre. He died in Paris on August 15, 1758.

Boyle, Robert (1627–91) This English natural philosopher discovered that the volume occupied by a gas is inversely proportional to the pressure under which it is held (known in English-speaking countries as Boyle's law) and that the weight of a body varies according to the amount of buoyancy supplied by the atmosphere. Boyle was born on January 25, 1627, at Lismore

Castle, Ireland. He was the seventh son of the earl of Cork. After studying at Eton College and traveling in Europe, from 1644 Boyle devoted himself to a life of scientific inquiry. He moved to Oxford in 1654 and it was in that city that he carried out his most important scientific work. Helped by Robert Hooke, Boyle studied combustion, discovering that neither charcoal nor sulfur will burn when air is excluded, but they burn readily when air is allowed into the container. When either substance is mixed with potassium nitrate (saltpeter), however, the mixture will burn in a vacuum. Boyle concluded from this that both potassium nitrate and air contain some ingredient necessary for combustion. The ingredient is oxygen, but Boyle failed to identify it. Boyle was the first scientist to use the word "analysis" to describe the separation of a substance into its constituents. He invented a hydrometer for measuring the density of liquids and made the first match by coating a rough paper with phosphorus and placing a drop of sulfur on the tip of a small stick that ignited when drawn along a crease in the paper. He made a portable camera obscura that could be extended or shortened like a telescope in order to focus an image on a piece of paper stretched across the back of the box, opposite the lens. The charter establishing the Royal Society, granted by King Charles II and passed on August 13, 1662, named Boyle as a member of the council. In 1680 he was elected president, but declined because he was unwilling to take the necessary oath. In 1668 Boyle returned to live in London with his sister, where he remained for the rest of his life. He died on December 30, 1691.

Brückner, Eduard (1862–1927) The German geographer and glaciologist who studied climate change and identified the 35-year cycle that bears his name was born at Jena, Saxony, on July 29, 1862. He was educated at the gymnasium (high school) in Karlsruhe and from 1881 until 1885 he studied physics and meteorology at the University of Dorpat (now Tartu, Estonia). He then continued his studies in Dresden and Munich. In 1885 Brückner joined the staff of the Deutsche Seewarte in Hamburg. The Seewarte later developed into the modern German weather service. From 1888 until 1904 he was a professor at the University of Bern, Switzerland, and from 1899 to 1900 he was rector of the university. In 1904 he returned to Germany to become a professor at the University of Halle, and in 1906 he was appointed a professor at the University of Vienna, a post he held until his death. Brücker was convinced that climate change is of great importance, with direct economic and social implications, and published many scientific papers on the subject. He died in Vienna on May 20, 1927.

Buchan, Alexander (1829–1907) The Scottish scientist acknowledged to have been the most eminent British meteorologist of the 19th century was born at Kinnesswood, Kinross, Scotland, on April 11, 1829. Buchan established his reputation in 1867, when he published his *Handy Book of Meteorology.* This became a standard textbook and remained in use for many years. In 1869 he wrote a paper called "The Mean Pressure of the Atmosphere and the Prevailing Winds Over the Globe" for the Royal Society of Edinburgh. He also wrote papers on the circulation of the atmosphere and on ocean circulation. It was in 1869 that he published his paper on "Interruptions in the Regular Rise and Fall of Temperature in the Course of the Year" in the *Journal of the Scottish Meteorological Society,* describing what came to be called Buchan spells. In December 1860 Buchan was appointed secretary to the Scottish Meteorological Society, and he remained in the post until his death in 1907. From 1864, he also edited the *Journal of the Scottish Meteorological Society* from its first issue and wrote a great deal of its material. During his editorship, the journal published Thomas Stevenson's description of his louvered screen. Buchan was made a member of the Meteorological Council in 1887, and was elected a fellow of the Royal Society in 1898. In 1902 he was the first person to be awarded the Symons Medal, the greatest honor meteorologists can bestow on one of their colleagues. Alexander Buchan died in Edinburgh on May 13, 1907.

Budyko, Mikhail Ivanovich (b. 1920) Belorussian physicist and meteorologist who was the first scientist to calculate the balance of heat received from the Sun and radiated from the Earth's surface, checking his calculations against observational data from all parts of the world. He was born at Gomel, Belarus, and educated in Leningrad (now St. Petersburg); from

1942 until 1975 he worked at the Main Geophysical Observatory, Leningrad, where he was the director from 1972 to 1975. He was then appointed head of the Division for Climate Change Research, at the State Hydrological Institute, Leningrad. In 1956 he published his book *Heat Balance of the Earth's Surface.* This work changed climatology into a more physical discipline. Budyko became a pioneer of physical climatology. By 1960, he was already concerned about the possibility of climate change due to human activity. In 1972 he confirmed a link between past climate changes and changes in the atmospheric concentration of carbon dioxide. His analysis indicated a general warming of the world's climates caused by a rise in carbon dioxide concentration from the increasing consumption of fossil fuels. His calculations predicted a rise in temperature of about 6.3°F (3.5°C) between 1950 and about 2070. His studies of the effects on climate of altering the composition of the atmosphere led him, in the early 1980s, to ponder the climatic consequences of a large-scale thermonuclear war. He suggested that such a war might inject such a huge quantity of aerosols into the atmosphere that the entire world would be plunged into deep cold, a "nuclear winter" that might threaten human survival. Mikhail Budyko was elected an Academician of the Russian Academy of Sciences in 1992. He has been awarded many prizes, including the Lenin National Prize (1958), Gold Medal of the World Meteorological Organization (1987), A.A. Grigoryev Prize of the Russian Academy of Sciences (1995), and the Blue Planet Prize (1999) for his contribution to environmental research.

Buys Ballot, Christoph Hendrick Diderik (1817–90) The Dutch meteorologist who proposed the law bearing his name was born at Kloetinge, Zeeland, on October 10, 1817. He was educated at the University of Utrecht, obtaining his Ph.D. in 1844, and in 1847 he was appointed professor of mathematics at the University of Utrecht. In 1854 he helped to found and was the first director of the Royal Netherlands Meteorological Institute. He remained in this post until his death. In 1857 Buys Ballot described the wind circulation around areas of low and high atmospheric pressure. He based his description on his studies of meteorological records and it quickly became known as a law attributed to him although, unknown to Buys Ballot,

William Ferrel had reached the same conclusion on theoretical grounds some months earlier. When he learned of this, Buys Ballot acknowledged Ferrel's prior claim to the discovery, but it was too late to change the popular name of the law. Buys Ballot died at Utrecht on February 3, 1890.

Cavendish, Henry (1731–1810) The English physicist who discovered that the composition of air is the same everywhere and at all times was born on October 31, 1731, at Nice, France. His paternal grandfather was the duke of Devonshire and his maternal grandfather was the duke of Kent; Cavendish was very wealthy. He was educated at Cambridge University but left without taking a degree. He studied "fixed air" (carbon dioxide), "inflammable air" (hydrogen), and "common" (i.e., atmospheric) air. In 1783, after taking samples on 60 days and making 400 analyses of them, he found not only that air is everywhere and always the same, but also that a small proportion of it is inert. This was later found to be mainly argon. He also studied heat, and discovered latent heat and specific heat before Joseph Black, but did not publish his results. His work on electricity included sending sparks through the air and thereby oxidizing nitrogen. When he dissolved the resulting gas in water and analysed it, he found it was nitric acid. He determined the freezing point of mercury. He died at Clapham, London, on February 24, 1810.

Celsius, Anders (1701–44) The Swedish astronomer and physicist, who gave his name to the Celsius temperature scale, was born in Uppsala, Sweden, on November 27, 1701, into a family of eminent scientists. He was educated in Uppsala and in 1730 succeeded his father as professor of astronomy at the University of Uppsala. It was Celsius and Olof Hiorter, his assistant, who discovered that the aurorae are magnetic phenomena. His success as an astronomer persuaded the Swedish government to finance the building of an observatory at Uppsala. The Celsius Observatory opened in 1741, with Celsius as its first director. In 1742 Celsius proposed that all scientific measurements of temperature should be made on a scale based on two fixed points that occur naturally. This led to the development of the temperature scale that bears his name. Celsius published most of his scientific papers through the

Royal Swedish Academy of Sciences and he was its secretary from 1725 until 1744. Celsius died in Uppsala on April 25, 1744.

Charles, Jacques Alexandre César (1746–1823) The French physicist, mathematician, and pioneer balloonist, who discovered the law named after him, was born at Beaugency, Loiret, on November 12, 1746. In June 1783, the Montgolfier brothers made their first experiments with unmanned hot-air balloons at Annonay, in the south of France. The Academy of Sciences asked Charles to study the invention, and with the help of Nicolas and Anne-Jean Robert he successfully launched a hydrogen balloon in August 1783. On December 1, Charles and Nicolas Robert became the first people to ascend in a balloon. In later flights Charles reached a height of nearly 10,000 feet (3,000 m). Charles made his most important discovery in about 1787. Using oxygen, nitrogen, and hydrogen, he found that for every 1°C (1.8°F) rise in temperature their volume increased by 1/273 of the volume they had at 0°C (32°F). This meant that if the gas could be cooled to –273°C (–459.4°F) its volume would be zero. This came to be known as absolute zero. Charles did not publish the results of these experiments, but he did inform Joseph Gay-Lussac about them. Gay-Lussac repeated them and the resulting general rule came to be known in France as Gay-Lussac's law, but outside France it is called Charles's law. In 1785 Charles was elected to the Academy of Sciences. Later he became professor of physics at the Paris Conservatoire des Arts et Métiers. He died in Paris on April 7, 1823.

Clarke, Arthur Charles (b. 1917) The English science-fiction writer, who first proposed the use of satellites for communications and suggested placing satellites in geostationary orbit, was born at Minehead, Somerset, on December 16, 1917. During World War II Clarke served as an officer in the Royal Air Force, where he worked on the experimental trials of ground-controlled approach (GCA), the first use of radar to allow aircraft to land safely in poor visibility. He graduated in physics and mathematics from Kings College, London, in 1948. The technical paper in which he outlined the principles of geostationary orbits and the use of satellites for communications was published in 1945. In 1954, in a letter to Harry Wexler, head of the Scientific Services Division of the U.S. Weather Bureau, Clarke suggested using satellites to obtain data for weather forecasting. Arthur C. Clarke has received many awards and honors, including the 1982 Marconi International Fellowship, the gold medal of the Franklin Institute, the Lindbergh Award, the Vikram Sarabhai Professorship of the Physical Research Laboratory, Ahmedabad, and a fellowship of Kings College, London. He received a knighthood on May 16, 2000. Since 1956 Sir Arthur has lived in Colombo, Sri Lanka.

Coriolis, Gustave de (1792–1843) This French physicist, mathematician, and engineer was the first person to explain why moving bodies are deflected to the right in the Northern Hemisphere and to the left in the Southern Hemisphere. Coriolis was born in Paris on May 21, 1792; in 1808 he commenced his studies at the École Polytechnique, the school that trained government officials. He completed them at the École des Ponts et Chaussées, graduating in highway engineering. In 1816 he joined the staff of the École Polytechnique, first as a tutor and then as an assistant professor of analysis and mechanics. In 1829 he became professor of mechanics at the École Centrale des Arts et Manufactures, where he remained until 1836, when he became professor of mechanics at the École des Ponts et Chaussées. In 1838 he was made director of studies at the École Polytechnique. He was elected a member of the mechanics section of the Academy of Sciences in 1836. In 1835 he published a paper, called "Sur les équations du mouvement relatif des systèmes de corps" ("On the equations of relative motion of a system of bodies"), in the *Journal de l'École Polytechnique.* This showed that when a body moves in a rotating frame of reference a force of inertia deflects the body to the right if the frame of reference is rotating counterclockwise and to the left if the rotation is clockwise. This is now known as the Coriolis effect. Coriolis also established "work" as a technical term, defining it as the displacement of a force through a distance. In dynamics he introduced a quantity, $\frac{1}{2} mv^2$, for which he coined the term *force vive,* now called "kinetic energy." Coriolis died in Paris on September 19, 1843.

Croll, James (1821–90) The Scottish climatologist and geologist who proposed that ice ages occur at intervals of 100,000 years; born near Cargill, Perthshire, on January 2, 1821. A self-educated man from a poor family, Croll left school at 13 and had a number of jobs before publishing a book called *The Philosophy of Theism* in 1857. The book attracted some attention and led to his appointment as keeper of the Andersonian Museum in Glasgow in 1859. This gave him access to the library, where he continued his self-education. He published several papers on chemistry, physics, and geology, and in 1867 he was placed in charge of the Edinburgh office of the Geological Survey of Scotland, a post he held until he retired in 1880. In 1864 Croll proposed that the onset of glacial periods is triggered by changes in the eccentricity of the Earth's orbit and the precession of the equinoxes. This helped to prepare the way for the theory of Milankovich cycles. He published several books, including *Climate and Time, in Their Geological Relations* (1875), *Discussions on Climate and Cosmology* (1885), *Stellar Evolution and Its Relations to Geological Time* (1889), and *The Philosophical Basis of Evolution* (1890). Croll died from heart disease near Perth on December 15, 1890.

Crutzen, Paul (b. 1933) This Dutch atmospheric chemist, who shared the 1995 Nobel Prize for chemistry for his part in the discovery of the processes that deplete the ozone layer, was born on December 3, 1933, in Amsterdam. The Netherlands was invaded and occupied by German forces in May 1940, and Crutzen's education began during the occupation period and was interrupted several times. In 1951 he enrolled at a technical school for a three-year course in civil engineering, graduating in 1954, after which he worked for the Bridge Construction Bureau of the city of Amsterdam from 1954 until 1958, with an interruption for his compulsory military service from 1956 to 1958. Despite knowing nothing at all about computer programming, in 1958 Crutzen successfully applied for a post as a computer programmer in the Department of Meteorology of Stockholms Högskola (which became Stockholm University in 1960). The Department of Meteorology (now the Meteorology Institute) and the International Meteorological Institute associated with it were at the forefront of research and housed some of the fastest computers in the world. Until 1966, Crutzen spent much of his time building and running some of the first weather prediction models. Because he worked at the university, Crutzen was able to attend some of the lectures. In 1963 he obtained his master of science degree in mathematics, statistics, and meteorology, in 1968 his Ph.D. in stratospheric chemistry, and his D.Sc. (Doctor of Science, which is a higher degree than a Ph.D.) in 1973, for research into the photochemistry of ozone. Since then Crutzen has held several important posts. He was an adjunct professor in the Atmospheric Sciences Department of the University of Colorado from 1976 to 1981. In 1980 he was appointed director of the Atmospheric Chemistry Division of the Max-Planck Institute for Chemistry, at Mainz, Germany, and from 1983 to 1985 he was executive director of the institute. He held a part-time professorship at the University of Chicago from 1987 until 1991, and since 1992 he has been a part-time professor at the Scripps Institution of Oceanography of the University of California. Professor Crutzen has received many honors and holds honorary degrees from universities at York, Canada, Louvain, Belgium, East Anglia, England, and Thessaloniki, Greece.

Dalton, John (1766–1844) The English meteorologist and chemist who in 1803 proposed the law of partial pressures (Dalton's law) was born at Eaglesfield, near Cockermouth, Cumberland (now Cumbria), on September 6, 1766, the son of a weaver and the third of six children. Dalton also discovered that the density of water varies with its temperature and studied what happens when substances dissolve in water and when gases mix, concluding that water and gases must consist of very small particles that intermingle, so that the particles of a dissolved substance are located between water particles. In his book *New System of Chemical Philosophy,* published in 1808, he suggested that the particles of different elements have different weights. He compiled a list of these weights (relative atomic masses) and devised a system of symbols for the elements. These could be combined to represent compounds. From 1787, Dalton kept a diary of meteorological observations in which he made entries regularly for 57 years until his death. It finally contained more than 200,000 observations. He earned his living as a teacher, becoming a headmaster, and in his spare time he wrote articles on scientific topics for popular magazines. His book *Meteorological Observations and Essays* appeared in 1793, but did not sell well. In the same year he moved to Manchester to teach mathematics and natural philosophy at New College. He held this post until 1799, when the college was moved to York; Dalton remained in Manchester as a private teacher of mathematics and chemistry. Dalton became very famous. He delivered two courses of lectures at the Royal Institution in London, in 1804 and 1809–10. He was elected a fellow of the Royal Society in 1822. He became a corresponding member of the French Academy of Sciences and in 1830 he was elected one of its eight foreign associates. In 1833 the government awarded him an annual pension of £150 and raised it to £300 in 1836. John Dalton died in Manchester on July 27, 1844.

Dansgaard, Willi (b. 1922) Danish geophysicist who was the first scientist to show that measurements of the ratios of the oxygen isotopes ^{16}O and ^{18}O and deuterium in glacier ice could be used in the reconstruction of past climates. Dansgaard was born in 1922 in Copenhagen; he was educated at Copenhagen University, from where he obtained his Ph.D. in 1961. He is now emeritus professor of geophysics at Copenhagen University. His major work was done with the ice core drilled from the Greenland ice cap in 1966 by the U.S. Army. This core, 4,600 ft. (1,403 m) long, contained a climate record extending over 100,000 years. As well as devising a method for measuring the isotopes in the ice, Dansgaard and his colleagues measured the acidity of the ice and its dust content, both of which are climatically significant. Dansgaard was awarded the 1996 Tyler Prize for environmental achievement. He is a member of the Royal Danish Academy of Science and Letters, the Royal Swedish Academy of Sciences, the Icelandic Academy of Sciences, and the Danish Geophysical Society, and holds a number of major scientific awards and medals.

Deluc, Jean André (1727–1817) This Swiss geologist, meteorologist, and physicist made several important contributions to geology and meteorology, as well as inventing the dry pile electric battery. He was born in Geneva on February 8, 1727, and in

1773 he moved to England. Deluc discovered that water reaches its maximum density at 40°F (4°C). In 1761 he found that the heat required to melt ice or vaporize liquid water does not raise their temperatures. This was the concept of latent heat discovered independently at about the same time by Joseph Black. Deluc was also the first scientist to propose that the amount of water vapor that can be contained in a given space is independent of any other gases that may be present in that space. He invented a hygrometer, though not a very successful one, and he was the first person to devise a way of measuring height by means of a barometer. He showed that an increase in elevation is proportional to a decrease in the logarithm of the air pressure, and that a change in elevation is also inversely proportional to the air temperature. He was elected a fellow of the Royal Society and appointed a reader to Queen Charlotte, a post that brought him an income but did not make excessive demands on his time. Jean Deluc died at Windsor, England, on November 7, 1817.

Dobson, Gordon Miller Bourne (1889–1976) The English meteorologist who spent much of his career studying stratospheric ozone was born on February 25, 1889. During World War I he was director of the experimental department at the Royal Aircraft Establishment, Farnborough, and in 1920 he became a lecturer in meteorology at the University of Oxford. He studied meteor trails and deduced that there is a region of the stratosphere where the temperature increases with height. Dobson inferred that this warming was due to the absorption of ultraviolet solar radiation by ozone and determined to measure the concentration of stratospheric ozone. In order to do so, in 1924 he built a spectrograph in his workshop at home. The measurements he made with it during 1925 established the way the ozone concentration varies with the seasons and the relationship between the ozone concentration and weather conditions in the upper troposphere and lower stratosphere. He went on to build and calibrate five more spectrographs for use at other European locations. In 1927 or 1928 Dobson built a spectrophotometer that was more sensitive than his earlier instruments. He became interested in atmospheric pollution early in the 1930s and helped develop reliable methods for measuring smoke, sulfur dioxide, and other pollutants. During

World War II Dobson studied stratospheric humidity to devise a means of forecasting the height at which aircraft condensation trails would form. In the course of this work he designed the first frost-point hygrometer. After the war he returned to the study of ozone. By 1957 there were 44 Dobson spectrometers in use in various parts of the world. There are now more than 150 in regular use and ozone concentration is reported in Dobson units. Dobson became a fellow of the Royal Society in 1927 and in 1945 Oxford University conferred the title of professor on him. He retired from the university in 1956, but continued to study ozone. His last observation was made the day before he suffered the stroke from which he died on March 11, 1976, at Oxford.

Doppler, Johann Christian (1803–53) The Austrian physicist who discovered the effect that bears his name was born on November 29, 1803, in Salzburg and educated there and from 1822 to 1825 at the Polytechnic Institute in Vienna, where he studied physics and mathematics. The Doppler effect was first tested in 1845 at Utrecht, in the Netherlands. In 1835 Doppler was appointed professor of mathematics at the State Secondary School in Prague and subsequently held professorships at the State Technical Academy, Prague, and the Mining Academy, Schemnitz. In 1850 he returned to Vienna as director of the Physical Institute and professor of experimental physics at the Royal Imperial University of Vienna. He died in Venice, Italy, on March 17, 1853.

Ekman, Vagn Walfrid (1874–1954) A Swedish oceanographer and physicist who was born in Stockholm on May 3, 1874, and educated at the University of Uppsala, where he received his doctoral degree in 1902, Ekman discovered that the movement of wind-driven ocean currents is affected by friction between different layers of water and by the Coriolis effect. Together these produce a change in the direction of currents with depth that became known as the Ekman spiral. A similar effect occurs in the atmosphere. After receiving his doctorate Ekman moved to Norway (which was still part of Sweden) to take up a post as an assistant at the International Laboratory for Oceanographic Research in Oslo (then called Kristiania). He remained there until 1908 and returned to

Sweden in 1910, as professor of mechanics and mathematical physics at the University of Lund, a post he held until he retired in 1939. Ekman died on March 9, 1954, at Gostad, near Stockaryd, Sweden.

Elsasser, Walter Maurice (1904–91) The German-born American physicist, who developed the theory that the Earth's core acts as a dynamo, generating a magnetic field, was born at Mannheim, Germany, on March 20, 1904. He was educated at the University of Göttingen, where he obtained his doctorate in 1927. He left Germany in 1933 and spent three years in Paris before moving to the United States. In 1940 he became an American citizen. In the course of his career Elsasser was professor of physics at the University of Pennsylvania, professor of geophysics at Princeton University, and a research professor at the University of Maryland. Elsasser also pioneered the analysis of the magnetic field recorded in the orientation of rock particles. This has been of great importance in tracing the movements of continents and the history of climate. Elsasser died on October 14, 1991.

Fabry, Marie Paul Auguste Charles (1867–1945) A French physicist who in 1896, with his colleague the French physicist Albert Pérot (1863–1925), invented the Fabry–Pérot interferometer and Fabry–Pérot etalon. Using these instruments, Fabry and Pérot were able to confirm the Doppler effect for light in the laboratory and they applied the instruments to a variety of astronomical questions. In 1913 Fabry used the interferometer to discover that ultraviolet radiation was being absorbed by ozone, which he found was abundant in the upper atmosphere. Fabry was born in Marseilles on June 11, 1867, and was educated in Marseilles and at the École Polytechnique in Paris, from where he graduated in 1889 in physics and mathematics. After graduating, Fabry moved to the University of Paris, where he obtained his doctorate in physics in 1892. He then spent two years teaching physics at lycées (high schools) in several cities before joining the staff at the University of Marseilles in 1894, where he devoted himself to teaching and pursuing his own research. In 1904 he was appointed professor of industrial physics. In 1914 the French government called him to Paris to investigate interference in sound and light waves, and in 1921 Fabry moved to Paris permanently as professor of physics at the Sorbonne. Later he combined this post with that of professor of physics at the École Polytechnique and director of the Institute of Optics. In 1935 he became a member of the International Committee on Weights and Measures. He retired two years later, in 1937, and died in Paris on December 11, 1945.

Fahrenheit, Daniel Gabriel or **Gabriel Daniel** (1686–1736) The Polish–Dutch physicist who devised the temperature scale bearing his name was born on May 14, 1686, in Danzig (now Gdansk). He began his education in Danzig, but moved to Amsterdam in 1701 in order to learn a business. He became interested in the making of scientific instruments and in about 1707 he left the Netherlands to tour Europe. In the course of his travels, in 1708 he met the Danish physicist and instrument maker Claus Römer (1644–1710), who had made and calibrated a thermometer using alcohol. Fahrenheit returned to Amsterdam, where in 1717 he set up his own instrument-making business, and remained there for the rest of his life. There was intense scientific interest in studying the atmosphere and weather, but meteorologists were greatly hindered by the lack of a reliable thermometer. Fahrenheit turned his attention to the problem and finally succeeded in making a satisfactory thermometer using mercury. He based his thermometer calibration on the scale he had seen Römer use, with two fiducial points. After some later adjustments, he produced the Fahrenheit temperature scale, still in use today, on which ice melts at 32° and water boils at 212°. Fahrenheit described his method for making thermometers in 1724, in a paper he submitted for publication in the *Philosophical Transactions of the Royal Society.* He was elected to the Royal Society in the same year. Fahrenheit died at The Hague on September 16, 1736.

Ferdinand II (1610–70) An Italian physicist who was a member of the powerful Medici family, Ferdinand was born on July 14, 1610, the son of Cosimo II, Grand Duke of Tuscany. Ferdinand took a keen interest in science, especially atmospheric science. One of the challenges of the time was to find a way to measure temperature, and in 1641 Ferdinand invented a thermometer consisting of a tube that contained liquid and was sealed at one end. He improved on it with a further design in 1654, his thermometer providing the basis for the instrument that would be made by Gabriel Fahrenheit about 60 years later. Ferdinand also designed one of the earliest accurate hygrometers. In 1657, the year he produced his hygrometer, Ferdinand and his brother Leopold founded the Accademia del Cimento (Academy of Experiments) in Florence. Members of the Accademia were especially interested in studying the atmosphere. Carlo Renaldini was one of those who worked on developing the thermometer. The Accademia itself was the forerunner of other scientific academies, including the Royal Society of London (founded in 1665) and the Royal French Academy of Sciences (founded in 1666). The Accademia ceased to function in 1667. Ferdinand died on May 24, 1670.

Ferrel, William (1817–91) The American climatologist who discovered that winds flow parallel to isobars was born on January 29, 1817, in Bedford County, Pennsylvania. In 1856 he published a mathematical model of the circulation of the atmosphere, revising it in 1860 and 1889. The model proposed the existence of an indirect midlatitude cell, now called the Ferrel cell. In 1857 his particular interest in and understanding of tides led to an invitation to join the staff of *The American Ephemeris and Nautical Almanac*. While working for this publication Ferrel calculated that the combined effect of the pressure-gradient force and Coriolis effect must cause winds to blow at 90° to the pressure gradient, parallel to the isobars rather than across them. A few months later the Dutch meteorologist C. H. D. Buys Ballot announced the same discovery; although he acknowledged Ferrel's prior claim, it became known as Buys Ballot's law. In 1867 Ferrel joined the United States Coast and Geodetic Survey to develop a general theory of the tides and in 1882 moved to the Army Signal Service, where he continued to work until he retired in 1886. The Coast Survey asked him to complete the investigations on which he was engaged at the time he left and to continue supervising the design and construction of the tide-predicting machine he had invented. This was a mechanical device that predicted times and heights of high and low water. It was first used to predict the tides for 1885 and it remained in use until

1991, when computers took over the task. Ferrel wrote extensively on meteorology. His titles include *Meteorological Researches,* published in three volumes between 1877 and 1882, *Popular Essays on the Movements of the Atmosphere* (1882), *Temperature of the Atmosphere and the Earth's Surface* (1884), *Recent Advances in Meteorology* (1886), and *A Popular Treatise on the Winds* (1889). After his retirement William Ferrel moved to Maywood, Kansas, where he died on September 18, 1891.

FitzRoy, Robert (1805–65) The English naval officer, hydrographer, and meteorologist, who in 1860 prepared the world's first daily weather forecast published by a newspaper (the *Times* of London), was born on July 5, 1805, at Ampton Hall, in Suffolk. He was educated at the Royal Naval College, Portsmouth, entered the Royal Navy after graduating in 1819, and received his commission as an officer on September 7, 1824. He served in the Mediterranean and was then sent to South America on HMS *Beagle,* which was conducting a surveying mission. When the captain of the *Beagle* died FitzRoy assumed command, completing the survey and returning to England. He applied to lead a second survey and in 1831 the Naval Hydrographer, Admiral Sir Francis Beaufort, granted the request. FitzRoy sailed once more as captain of the *Beagle,* this time accompanied by Charles Darwin. The *Beagle* carried several barometers. FitzRoy used these to prepare short-term weather forecasts. This was also the first voyage in which wind observations were based on the Beaufort wind scale. In 1841, by then an admiral, he became a member of Parliament for Durham and in 1843 he was made governor-general of New Zealand. At the insistence of the British settlers he was recalled from New Zealand in 1845, mainly because he believed the Maori claims to land were as legitimate as theirs. Admiral FitzRoy retired from active service in 1850. In 1854 he was appointed to head the newly formed meteorological department at the Board of Trade, where he encouraged the collection of weather observations, established barometer stations, and used telegraphy to gather data. These allowed the department to issue weather forecasts and, in 1861, the first storm warnings. In 1863 he published *The Weather Book* in which he set out principles to guide sailors in forecasting the weather. FitzRoy believed a barometer should be installed at every port. "FitzRoy" barometers became very popular and FitzRoy himself invented some versions. Robert FitzRoy died on April 30, 1865, at Upper Norwood, near London.

Flohn, Hermann (1912–97) The German meteorologist and climatologist, who devised a widely used climate classification, was born at Frankfurt-am-Main on February 19, 1912. He was educated at the Universities of Frankfurt and Innsbruck, graduating from both in meteorology, geography, geophysics, and geology. After working for a few months at the University of Marburg, in 1935 he joined the state weather service in Berlin. During the war he served in the Luftwaffe (air force) meteorological service and became a prisoner of war. After the war he joined the German weather service at Bad Kissingen. From 1952 to 1961 he was head of research at the weather service. From 1961 to 1977 he was a professor at the University of Bonn and director of the Meteorological Institute. After his retirement in 1977 he was made professor emeritus, but continued to head research projects. The last of these investigated large-scale climate forecasting and its environmental significance, and was conducted for the North Rhine–Westphalia Academy of Science. He died in Bonn on June 23, 1997.

Fourier, Jean-Baptiste Joseph (1768–1830) The French mathematician and physicist, who developed a mathematical description of heat, was born on March 21, 1768, at Auxerre, southeast of Paris. He was educated at the Auxerre military academy and at a Benedictine school in St. Bênoit-sur-Loire. He returned to Auxerre in 1784 and taught mathematics at the military academy. The École Normale opened in Paris in 1795 and Fourier taught there, acquiring such a reputation that before long he was made professor of analysis at the École Polytechnique. In 1798 he was chosen to accompany Napoleon on his campaign in Egypt. He was made governor of Lower Egypt and remained there until 1801. On his return to France he was made prefect of Isère. In 1808 Napoleon conferred the title of baron on him and later made him a count. Fourier rejoined Napoleon in 1815, and after Napoleon's final defeat he settled the following year in Paris. Fourier was elected to the Academy of Sciences in 1817. He was also elected to the Académie Française and to foreign membership of the Royal Society of London. Fourier explained his theory of heat in *Théorie Analytique de la Chaleur.* Published in 1822 and translated into English *(Analytical Theory of Heat)* in 1872, this proved to be one of the most influential scientific books of the 19th century. In connection with it, Fourier developed a technique that allows the overall description of a complex phenomenon to be broken down into a series of simpler, trigonometric equations, known as the Fourier series, the sum of which is equal to the original description. The Fourier series can be applied to any complex function that repeats and so it is of value in many branches of physics. It is widely used by meteorologists. He also developed the use of linear partial differential equations for solving boundary-value problems. This is relevant to numerical forecasting. He also investigated probability theory and the theory of errors. Fourier died in Paris on May 16, 1830.

Franklin, Benjamin (1706–90) American statesman, physicist, inventor, author, and publisher who demonstrated that lightning is an electrical discharge; born on January 17, 1706, in Boston, Massachusetts. He was educated privately, but was mainly self-taught. His greatest scientific achievements arose from his study of electricity. Like many other scientists, Franklin experimented with a Leyden jar. This device stored a static electrical charge and had a projecting metal rod from which a spark would travel, with a loud crackling noise, to any other metal object held close to it. His observations led him to wonder whether the spark and crackle might not be a tiny demonstration of lightning and thunder and, therefore, whether during a thunderstorm the sky and Earth might become a giant Leyden jar. To test the idea, in 1752 he flew a kite in a thunderstorm. He had attached a pointed piece of wire to the kite and tied a long silken thread to the wire. At the bottom of the thread he tied a metal key. As the kite flew near the base of the storm cloud and lightning began flashing nearby, Franklin held his hand close to the key and a spark jumped to it. Then he held a Leyden jar to the key and it accumulated an electric charge. He had proved that lightning is a discharge of electricity and thunder is the sound of the spark. That same year the French scientist Thomas-François d'Alibard (1703–99) also proved, independently of Franklin, that lightning is an electrical phenomenon. Franklin also suggested that buildings could be protected from damage by lightning if pointed metal rods were fixed to their roofs and connected to the ground by metal wires. He had invented the lightning conductor. Franklin died at Philadelphia on April 17, 1790.

Fujita, Tetsuya Theodore (1920–98) The Japanese-American meteorologist who became one of the world's leading authorities on tornadoes was born in Kitakyushu City, Japan, on October 23, 1920. He was a student specializing in the study of tropical cyclones when the atomic bomb was dropped on Hiroshima; he became intensely interested in the tornadoes that were triggered by the firestorms caused by the bomb. He moved to the United States in the early 1950s and continued his investigations of tornadoes at the University of Chicago. Fujita remained at the University of Chicago for the rest of his life, becoming professor of meteorology. He adopted the name Theodore in 1968, and was known to his friends as Ted. Fujita introduced the concept of tornado families, a number of tornadoes triggered by a single storm, and in the 1970s he and his wife Sumiko devised the tornado intensity scale that bears their name. He also identified two other phenomena, the downburst and microburst. Toward the end of his career he returned to the study of tropical cyclones. Professor Fujita died at Chicago on November 19, 1998.

Galileo Galilei (1564–1642) The Italian physicist and astronomer, and one of the most famous scientists who ever lived, was born at Pisa on February 15, 1564. He is usually identified by his given name, Galileo, rather than by his family name, Galilei. He was educated by a private tutor, then at a monastery in Vallombrosa, near Florence, and finally at the University of Pisa, where he studied medicine but did not take a degree. Following his return to Florence he obtained a post as a lecturer in mathematics and science at the Florentine Academy. In 1589 Galileo became professor of mathematics at the University of Pisa, and in 1592 he applied for and was awarded the better paying post of professor of mathematics at the University of Padua. He remained at Padua for the next 18 years and it is there that he did most of his best work. In 1593 Galileo invented the first thermometer, called a thermoscope. It was highly inaccurate, because no account was taken of changes in atmospheric pressure, but it was one of the earliest attempts to construct an instrument for making scientific measurements. Toward the end of his life, Galileo became interested in discovering whether air is a physical substance having mass. A young assistant, Evangelista Torricelli, set to work on the problem, and the experimental apparatus he devised was the first barometer. Galileo died from a fever at Arcetri, near Florence, on January 8, 1642.

Galton, Francis (1822–1911) An English scientist, inventor, and explorer, born on February 16, 1822, near Sparkbrook, now a suburb of Birmingham, Galton was the first scientist to plot meteorological data onto a map and to attempt to produce a synoptic chart showing the weather conditions over a large area. He was the son of a wealthy banker and a first cousin of Charles Darwin. A prodigy, Galton was able to read before he was three years old and was studying Latin by the time he was four. In response to his father's wishes he studied medicine at Birmingham General Hospital and at King's College, London, but interrupted his medical studies to study mathematics at Trinity College, Cambridge. He then resumed studying medicine at St. George's Hospital, London, but never completed the course. His father died and Francis inherited a fortune, so he left the hospital and for the rest of his life he pursued whatever topic interested him. He began by traveling extensively. His observations in what was then a little known part of Africa led to the award of the Gold Medal of the Royal Geographical Society in 1854 and in 1856 he was elected a Fellow of the Royal Society. In the 1860s Galton began to study the weather and in 1862 he finally succeeded in compiling a detailed weather map. Other meteorologists had already established the cyclonic circulation of air around areas of low pressure. Galton discovered anticyclonic circulation around a center of high pressure and he coined the term anticyclone. He published the results of his research in 1863 in a monograph called *Meteorographica* (published by Macmillan) and summarized them, much later, in his book

Memories of My Life, published in 1908 by Methuen. He played a large part in preparing the daily weather charts that were published by the *Times* of London from data supplied by the Meteorological Office. As well as preparing weather charts for publication, Galton helped devise a way to print them using movable type. Galton had the idea of measuring the speed and direction of the wind at a specified location and time by means of the smoke emitted by an exploding shell and, on the suggestion of Robert FitzRoy, he also invented the wind rose. In addition to his contributions to meteorology, Francis Galton was the first person to demonstrate the uniqueness of fingerprints and partly worked out a system for identifying them. He invented a teletype printer and the ultrasonic dog whistle, and also devised new techniques for statistical analysis, as well as a word-association test that was adopted by Sigmund Freud (1856–1939). Francis Galton was knighted in 1909. He died at Haslemere, Surrey, on January 17, 1911.

Gay-Lussac, Joseph Louis (1778–1850) The French chemist and physicist who was one of the founders of the science of meteorology was born on December 6, 1778, at St. Léonard, Haute Vienne, in central France. He was educated locally and at the École Polytechnique and École des Ponts et Chaussées (School of Bridges and Roads), graduating in 1800. In 1802 Gay-Lussac was appointed a demonstrator in chemistry at the École Polytechnique and in 1810 he became professor of chemistry. He was also professor of physics at the Sorbonne University in Paris from 1808 until 1832, when he resigned to take up the post of professor of chemistry at the Musée National d'Histoire Naturelle. In 1806 he was made a member of the Academy of Sciences. He published his first research results in 1802. In collaboration with Louis Jacques Thénard (1777–1857), Gay-Lussac had formulated a law stating that all gases expand by the same fraction of their volume when the temperature is increased by a given amount. Jacques Charles had discovered this law in 1787, but had not published it. On September 16, 1804, Gay-Lussac made a solo balloon ascent to 23,012 feet (7,019 m), establishing an altitude record that stood for 50 years. Measurements he made during this and a previous ascent showed that the chemical composition of the atmosphere remains constant with height. In 1809 Gay-Lussac published his discovery that when gases combine they do so in simple proportions by volume and that the products of their combination are related to the original volumes. This is known as Gay-Lussac's law and it is used in chemical equations. One of the examples Gay-Lussac used to illustrate it shows that when two molecules of carbon monoxide (CO) combine with one molecule of oxygen (O_2), the product is two molecules of carbon dioxide (CO_2). His advice was constantly in demand and he held a number of official positions. In 1805 he was appointed to the consultative committee on arts and manufactures. In 1818 he was appointed to the department responsible for the manufacture of gunpowder, and in 1829 he became chief assayer to the mint. Joseph Gay-Lussac died in Paris on May 9, 1850.

Hadley, George (1685–1768) English meteorologist who proposed the atmospheric circulation system that bears his name; born in London on February 12, 1685. He qualified in law, but became increasingly interested in physics, and atmospheric physics in particular. He was placed in charge of the meteorological observations that were prepared for the Royal Society, a task he performed for at least seven years. In 1686 Edmund Halley had proposed an explanation for the reliability of the trade winds that failed to account for their easterly direction. Hadley proposed that this is due to a deflection caused by the rotation of the Earth. Hadley produced the first model for the general circulation of the atmosphere. This supposed one large convection cell in each hemisphere. Warm air rises over the equator, moves to the pole where it subsides, and then returns to the equator. His contribution to meteorology is acknowledged to this day in the name of the Hadley cells. George Hadley died at Flitton, Bedfordshire, on June 28, 1768.

Halley, Edmund (1656–1742) This English astronomer proposed the first plausible explanation for the trade winds; he was born at Haggerston, Shoreditch, then a village near London, on October 29 according to the Julian calendar then in use (November 8 according to the Gregorian calendar in use today), 1656. He was educated at St. Paul's School, London, and Queen's College, University of Oxford, where he wrote a book on the laws of Johannes Kepler. Halley left Oxford without taking a degree, a practice that was not unusual, and in 1678 Oxford University awarded him a degree without requiring him to take the examination. By then he was a distinguished astronomer. Although astronomy was his principal interest, Halley also studied tides, winds, and weather phenomena. In 1686 he attempted an explanation of the trade winds by proposing that air is heated more strongly at the equator than elsewhere. The warm air rises and draws in cooler air from higher latitudes, flowing toward the equator. In 1678 Halley was elected to the Royal Society. In 1704 he was appointed Savilian Professor of Geometry at Oxford, and in 1720 he was appointed Astronomer Royal. Edmund Halley died in Greenwich, London, on January 14, 1742, by the Julian calendar (January 25 by the Gregorian calendar). His tombstone bears an inscription stating that he died in 1741, because in 17th-century England the year began on March 25.

Helmont, Jan Baptista van (1577–1644) The Flemish physician and alchemist, who discovered the existence of gases and claimed to have coined the word "gas," was born in Brussels (his date of birth is not known). He was educated at Louvain, graduating in medicine in 1599. In 1609 he moved to Vilvorde, near Brussels, where he spent the rest of his life practicing medicine and conducting chemical experiments. Some of his experiments produced vapors and he was the first person to recognize that these are distinct substances, each with its own properties. Unlike liquids and solids, vapors immediately fill any space they enter. He thought this meant they existed in a state of chaos, a word he spelled "gas," which is the way it sounded when spoken by a Flemish speaker. Charcoal gives off a gas when it is burned and because the gas comes from wood (Latin, *silva*) he called it "gas sylvestre." He discovered that the same gas is given off when malted barley is fermented to make beer. It is the gas we call carbon dioxide. Jan van Helmont continued experimenting and working as a physician until his death, on December 30, 1644, at Vilvorde.

Henry, Joseph (1797–1878) American physicist who invented the telegraph and used it to collect meteorological observations; born in Albany, New York, on December 17, 1797. He was educated at Albany Academy, where he studied chemistry, anatomy, and physiology, paying his way by teaching in country schools and tutoring. In 1825 he obtained a job surveying a route for a new road in New York State, but returned to Albany Academy in 1826 to teach mathematics and science. He invented the telegraph in 1831 and in 1832 he was appointed a professor at the College of New Jersey (now Princeton University). In December 1846 Joseph Henry was elected the first secretary of the Smithsonian Institution. He was also the second president of the National Academy of Sciences and was active in organizing the American Association for the Advancement of Science and the Philosophical Society of Washington. As one of his first projects at the institution Henry established a corps of voluntary observers, located all over the United States, who used the telegraph to send meteorological reports to a central office at the institution. For the next 30 years Henry supported and encouraged this volunteer corps. Joseph Henry died in Washington on May 13, 1878.

Hero of Alexandria (*c.* 60 CE) An engineer who lived in Alexandria, Egypt, Hero is sometimes described as Greek and sometimes as Egyptian. He was a gifted mathematician and teacher, who founded a school, part of which was devoted to research. He devised systems of gears for lifting heavy objects, used a suction machine to lift water, and invented a machine consisting of a hollow sphere that was made to spin rapidly by steam from boiling water. Hero demonstrated that air is a material substance made from minute particles that move independently of one another and that it can be compressed. This idea had to wait 1,600 years before being developed by Robert Boyle and his contemporaries.

Hooke, Robert (1635–1703) An English physicist who was one of the most ingenious experimenters and instrument makers who have ever lived; born on July 18, 1635, at Freshwater, in the Isle of Wight. As a child Hooke was often ill and spent much of his time alone. He was educated at schools in Oxford and London, and at Christ Church College, University of Oxford. While a student he earned a living selling ideas to the owners of workshops where scientific instruments were made. He did not take a degree. At Oxford he fell in with a group of brilliant scientists, including Robert Boyle, who employed him as an assistant. The two became lifelong friends. In 1659 the group moved to London, where they formed a society that in 1662 became the Royal Society of London. In 1663 Hooke was elected a fellow of the Royal Society and was also awarded an M.A. degree by Oxford University. In 1664 he was made a lecturer in mechanics at the Royal Society, and in 1665 he was appointed professor of geometry at Gresham College. He remained in these two posts for the rest of his life. From 1677 until 1683 he was secretary to the Royal Society. Hooke invented the wheel barometer, which indicated the pressure by means of a needle on a dial. He suggested ways to apply barometer readings to weather forecasting. It was Hooke who first labeled a barometer with the words "change," "rain," "much rain," "stormy," "fair," "set fair," and "very dry." He designed, but did not make, a weather clock to record air temperature, pressure, rainfall, humidity, and wind speed on a rotating drum. He suggested that the freezing point of water be used as the zero reference point on thermometers. Hooke died in London on March 3, 1703.

Howard, Luke (1772–1864) English chemist, pharmacist, and meteorologist who devised the first practical system of classifying clouds; born in London on November 28, 1772, into a Quaker family. Howard was educated at Thomas Huntly's School, at Burford, Oxfordshire, a Quaker school, from 1780 until 1787. He was trained as a pharmacist (druggist) and when he grew up he earned his living as a manufacturer of chemicals. He was never a professional scientist (and never claimed to be). His interest in meteorology began in the summer of 1783, when he was 11 years old. He began to keep a record of his meteorological observations and maintained the habit for more than 30 years. He presented papers on "The Average Barometer" in 1800 and "Theories of Rain" in 1802 to meetings of the Askesian Society, a scientific society of which he was a founding member. In 1803 (some historians say it was in December 1802), Howard presented a paper to the Askesian Society called "On the Modification of Clouds." By "modification" he meant what we would call "classification," and in this paper he allotted Latin names to cloud types and proposed ways of combining the names in a binomial fashion. His classification attracted widespread attention and Howard became a celebrity, his fame increasing even more when all his meteorological papers up to that time were collected by Thomas Forster and published in 1813 as *Researches About Atmospheric Phaenomenae.* Wolfgang von Goethe even dedicated four poems to Howard. In 1806, Howard began a publication called the *Meteorological Register,* which appeared regularly over several years in the *Athenaeum Magazine.* In 1818–19, he published the first book ever written on urban climate: *The Climate of London,* in two volumes. In 1833 this appeared as an expanded second edition in three volumes. In it he made what is believed to be the first reference to what is now called a heat island, with temperature records to support it. A series of lectures he delivered in 1817 were published in 1837 as *Seven Lectures in Meteorology* and became the first textbook on the subject. His last book, *Barometrographia,* appeared in 1847. In 1821 Howard was made a Fellow of the Royal Society. He died in London in March 1864.

Humboldt, Friedrich Heinrich Alexander, baron von (1769–1859) German geologist, geophysicist, geographer who was one of the founders of biogeography and made important contributions to climatology and meteorology; born in Berlin on September 4, 1769. Following the death of his father in 1779 Humboldt was educated privately before enrolling at the University of Göttingen in 1789 to study science. In 1791 he became a student at the Freiburg Bergakademie (School of Mining), graduating in 1793 in geology. He was then appointed assessor and later director of mines in the Prussian principality of Bayreuth. Accompanied by the French botanist Aimé Bonpland (1773–1858), Humboldt spent the years from 1799 to 1804 exploring South America. While following a route that crossed the Andes, Humboldt noted the changes in vegetation at different elevations and recorded the decrease in air temperature with height. When they reached the Pacific coast Humboldt measured the temperature of the water offshore and discovered the existence of the cold Peru Current (sometimes called the Humboldt Current). He also investigated the cause of tropical storms. His discoveries supplied information other scientists used later to determine the processes involved in the weather systems of the middle latitudes. He described the journey in his major work, *Voyage de Humboldt et Bonpland,* which appeared in 30 volumes between 1805 and 1834. His many discoveries and his liberal opinions made Humboldt a celebrity, said to be the second most famous man in Europe, after Napoleon. He died in Berlin on May 6, 1859, at the age of 89 and was given a state funeral.

Kepler, Johannes (1571–1630) German astronomer and mathematician who calculated the orbits of the planets; born on December 27, 1571, at Weil der Statt, Württemberg. He was educated at schools in Weil, Leonberg, and Adelberg, before entering the University of Maulbronn, from which he graduated in theology in 1588. He enrolled at the University of Tübingen in 1589 to study philosophy, mathematics, and astronomy, obtaining his master's degree in 1591. In 1594 Kepler was appointed professor of mathematics at the University of Graz. A religious purge forced him to leave Graz in 1598. He spent a year in Prague, returned to Graz only to be expelled once more, and went back to Prague, where he became assistant to the aging Danish astronomer Tycho Brahe (1546–1601) in 1600. It was his meticulous study of Brahe's voluminous records that led him to formulate the laws of planetary motion. In 1611 he wrote the world's first science fiction story, called *Somnium,* about a man who travels to the Moon. It was not published until 1631, after Kepler's death. It was also in 1611 that Kepler published a description of snowflakes, called *A New Year's Gift, or On the Six-cornered Snowflake,* dedicated to his patron and friend Matthaüs Wackher. Kepler died at Regensburg (then called Ratisbon), Bavaria, on November 15, 1630.

Kirchhoff, Gustav Robert (1824–87) The German physicist who discovered blackbody radiation was born in Königsberg, Prussia (now Kaliningrad, Russia), on March 12, 1824. He was educated at the University of Königsberg and in 1854 became professor of physics at the University of Heidelberg, where he met Robert Wilhelm von Bunsen (1811–99), inventor of the laboratory burner named for him. Working together, Kirchhoff and Bunsen developed the spectroscope, with which they were able to analyze compounds by the light spectra they emitted when heated with the Bunsen burner. Kirchhoff also calculated that a perfect blackbody, absorbing all the radiation falling on it, would emit radiation at all wavelengths if it were heated to incandescence. Balfour Stewart reached the same conclusion independently at about the same time, but Kirchhoff is usually credited with the discovery. Kirchhoff died in Berlin on October 17, 1887.

Köppen, Wladimir Peter (1846–1940) The German meteorologist and climatologist who devised the system of climate classification that bears his name was born in St. Petersburg, Russia, on September 25, 1846, of German parents. After attending school in the Crimea, he studied at the Universities of Heidelberg and Leipzig. From 1872 to 1873 he was employed in the Russian meteorological service. In 1875 he returned to Germany to take up an appointment as chief of a new division of the Deutsche Seewarte, based in Hamburg, where his task was to establish a weather forecasting service covering northwestern Germany and the adjacent sea areas. In 1884 he published the first version of his map of climatic zones. He plotted these on an imaginary continent he called "Köppen'sche Rübe" ("Köppen's beet"). His climate classification appeared in full in 1918 and after several revisions the final version of it was published in 1936. Köppen coauthored *Die Klimate der Geologischen Vorzeit (The climates of the geological past),* published in 1924, with Alfred Wegener (his son-in-law), and wrote *Grundriss der Klimakunde (Outline of climate science),* which was published in 1931. In 1927 he entered into collaboration with Rudolf Geiger to produce a five-volume work, *Handbuch der Klimatologie (Handbook of climatology).* This was never completed, but several parts, three of them by Köppen, were published. Köppen died at Graz, Austria, on June 22, 1940.

Lamb, Hubert Horace (1913–97) This English climatologist was one of the first scientists to draw attention to the variability of climates and the social and economic consequences of climate change; by the end of his life he was considered possibly the greatest climatologist of the 20th century. Hubert Lamb was born in Bedford and educated at Oundle School, Northamptonshire, and Trinity College, University of Cambridge, where he studied natural sciences and geography, graduating in 1935. He joined the Meteorological Office staff in 1936. His career almost ended in 1940, when he refused to investigate the meteorology of poison gas. He was moved to the Irish Meteorological Office and in 1941 was placed in charge of preparing forecasts for transatlantic flights. Lamb was among the most skillful of weather forecasters and during his period there, transatlantic flights out of Ireland had a perfect safety record. He returned to the Meteorological Office in Britain in 1945. In 1946–47 he sailed as meteorologist on a Norwegian whaling expedition to Antarctica, and it was during this voyage that he came to realize the extent to which climate changes over time. He served as a weather forecaster in Germany from 1951 to 1952, and from 1952 until 1954 he worked in Malta. In 1954 he was placed in charge of the Climatology Division at the Meteorological Office. While there he undertook the first detailed study of the past climate records held at the office. He used these to trace ways in which the climate had changed since the middle of the 18th century and devised a classification system for British weather. He also developed a method for estimating the climatic effect of volcanic dust. In 1971 Lamb left the Meteorological Office to establish the Climatic Research Unit at the University of East Anglia, in Norwich. He remained director of the unit until his retirement in 1977. Hubert Lamb died in Norwich on June 28, 1997.

Langley, Samuel Pierpont (1834–1906) American astronomer and physicist who was the first scientist to calculate the amount of energy the Earth receives from the Sun and the proportion of that energy which is absorbed by the atmosphere; born at Roxbury, Massachusetts, on August 22, 1834. He was educated at Boston Latin school, but he did not go to college and so was largely self-educated. By 1865 he had attained a sufficiently high standard to be offered a post as an assistant at the Harvard University observatory. In 1866 he left to teach mathematics at the U.S. Naval Academy, at Annapolis, Maryland, and in 1867 he was appointed director of the Allegheny observatory in Pennsylvania and professor of physics and astronomy at the Western University of Pennsylvania. In 1887 he became secretary and then director of the Smithsonian Institution, in Washington, D.C., a post he retained until his death. In 1881 he climbed Mount Whitney, California, accompanied by the American astronomer James Edward Keeler (1857–1900). From the summit the two were able to measure the heat of the solar rays and compare this with the value they measured at sea level. The measurements were made with a bolometer, an instrument Langley had invented for the purpose of studying the solar spectrum. Langley died at Aiken, South Carolina, on February 27, 1906.

Laplace, Pierre Simon, marquis de (1749–1827) The French mathematician and astronomer, who pointed the way toward the concept of the conservation of energy, was born on March 28, 1749, at Beaumont-en-Auge, Normandy. A mathematical and scientific genius, he enrolled at the University of Caen when he was only 16 and graduated in mathematics after two years. When he was 18 Laplace was appointed professor of mathematics at the École Militaire in Paris. He collaborated with the French chemist Antoine Laurent Lavoisier (1743–94) and in 1780 they were able to show that the amount of heat needed to decompose a compound into its constituent elements is equal to the amount released when those elements combine to form the compound. This developed further the work of Joseph Black on latent heat and pointed the way toward the concept of the conservation of energy. Sixty years later this was to become the first law of thermodynamics. In 1812 Laplace published *Théorie Analytique des Probabilités* (Analytical theory of probabilities), developing this further in *Essai Philosophique* (Philosophical essay), published in 1814. This work gave the theory of probability its modern form and helped establish statistics as a branch of mathematics. Napoleon made Laplace a count, and when Louis XVIII was restored to the throne in 1814, he was made a marquis. Laplace was elected to the Academy of Sciences in 1785. In 1816 he was elected to the Academie Française and in 1817 he became its president. He died in Paris on March 5, 1827.

Lorenz, Edward Norton (b. 1917) American meteorologist who discovered that weather systems exhibit chaotic behavior; born on May 23, 1917, in West Hartford, Connecticut and educated at Dartmouth College, Harvard University, and the Massachusetts Institute of Technology (M.I.T.), graduating in mathematics. During World War II, Lorenz served as a weather forecaster in the U.S. Army Air Corps. After the war he continued as a meteorologist. He joined the M.I.T. faculty in 1946 and received the degree of doctor of science in meteorology from M.I.T. in 1948. He was professor of meteorology at M.I.T. from 1962 until 1981, since when he has been professor emeritus of meteorology. Lorenz sought ways to improve forecasts by programming a computer to trace changes in an atmospheric variable, such as temperature, over a long period. One day in 1961 he ran the program after feeding it the initial conditions printed out from an earlier run. To his surprise, the second run traced changes that were markedly different from those of the first. He discovered that the difference arose because in the first run he had used a value with six decimal places, but the printout had rounded this to three decimal places. He had assumed, wrongly, that such a small difference would be insignificant. Lorenz nicknamed this extreme sensitivity to initial conditions as "the butterfly effect." Its discovery led him to further investigation of the mathematical theory of chaos.

Lovelock, James Ephraim (b. 1919) The English atmospheric chemist who proposed the Gaia hypothesis was born on July 26, 1919, at Letchworth Garden City, Hertfordshire. After leaving school he worked in a laboratory until he had gained the qualifications he needed to become a full-time student. He enrolled at the University of Manchester, graduating in chemistry in 1941. He then joined the staff at the National Institute for Medical Research, where he remained until 1961, when he became professor of chemistry at Baylor University College of Medicine, in Houston, Texas. He received the degree of D.Sc. in biophysics in 1959, from the University of London. In 1957 Lovelock invented the electron capture detector, which was used to reveal the presence of residues of organochlorine insecticides such as DDT throughout the natural environment, a discovery that contributed to the emergence of the popular environmental movement in the late 1960s. Later it registered the presence of minute concentrations of CFCs (chlorofluorocarbon compounds) in the atmosphere. In the early 1960s Lovelock became a consultant at the Jet Propulsion Laboratory (J.P.L.) of the California Institute of

Technology, in Pasadena. It was then that he began to speculate about how life on another planet might be detected and recognized, reasoning that any living organism must alter its environment by removing substances (such as food) from it and adding substances (such as body wastes) to it. This led to his formulation of the Gaia hypothesis, according to which the organisms living on a planet maintain environmental conditions favorable to them. In 1964 Lovelock became a freelance research scientist and inventor.

Magnus, Olaus (1490–1557) A Swedish priest and naturalist who was the author of a book containing the earliest European drawings of ice crystals and snowflakes, Olaf Mansson was born at Linköping in October 1490. "Olaus Magnus" is a latinized version of his name. He attended a school in Linköping, and then he and his elder brother Johannes spent nearly seven years traveling around Europe together to complete their education. Magnus was ordained a priest in 1519. He was a vicar in Stockholm in 1520 and dean of Strengnäs Cathedral in 1522. He had a distinguished ecclesiastical career, becoming archbishop of Sweden and director of the religious house of St. Brigitta, in Rome. His drawings of ice crystals and snowflakes were contained in his *Historia de gentibus septentrionalibus.* This was published in 1555 and became very popular throughout Europe. There were many editions and translations. The first English translation, called *History of the Goths, Swedes and Vandals,* appeared in 1658. It remains one of the most important sources of information about life in Scandinavia in the early 16th century. Magnus died in Rome on August 1, 1557.

Mariotte, Edmé (*c.* 1620–84) The French physicist who discovered the relationship between the volume of a gas and the pressure under which it is contained was born at Dijon in about 1620 and spent most of his life there. He was ordained as a priest, but it was to reward his scientific work that he was appointed prior of the abbey of Saint-Martin-sous-Beaune, near Dijon. His interests were wide and he wrote on many scientific topics, including vision, color, and plants, but his most important work concerned the behavior of fluids. He wrote about freezing and in 1679 he wrote an article called "De la nature de l'air" ("On the nature of air"). In this, Mariotte reported that "The diminution of the volume of the air proceeds in proportion to the weights with which it is loaded." This is the law discovered earlier by Robert Boyle. Mariotte discovered it independently, but noticed something else. He found that air expands when it is heated and contracts when it is cooled, so that the relationship between pressure and volume remains true only so long as the temperature remains constant. Boyle had overlooked this. The relationship is known in English-speaking countries as Boyle's law and in French-speaking countries as Mariotte's law. Mariotte died in Paris on May 12, 1684.

Marum, Martinus van (1750–1837) The Dutch chemist who discovered ozone and carbon monoxide was born on March 20, 1750. He was director of the Dutch Society of Sciences and the Teyler Museum at Haarlem from 1804 until 1837. In 1785, while subjecting oxygen to electrical discharges, van Marum noted "the odor of electrical matter" and the accelerated oxidation of mercury. The odor he described was that of ozone. He corresponded with many of the leading scientists of the day, including Benjamin Franklin, Joseph Priestley, Allesandro Volta (1745–1827), and Antoine Laurent Lavoisier (1743–94). Van Marum died on December 26, 1837.

Maunder, Edward Walter (1851–1928) English solar astronomer who established the link between sunspot activity and climate; born on April 12, 1851, in London. He was educated at Kings College, London, and then went to work at a bank. In 1873 he joined the staff of the Royal Observatory, Greenwich, as a photographic and spectroscopic assistant. Maunder was given the job of photographing sunspots and measuring their areas and positions. While engaged in this, his attention was drawn to the work of the German astronomer Gustav Spörer, who had identified a period from 1400 to 1510, when very few sunspots were seen. Maunder began searching through old records at the observatory to see whether Spörer was correct and whether there were any other such periods. It was this search that led to his discovery of the Maunder minimum, the period from 1645 to 1715 during which the recorded number of sunspots and auroras was extremely low. This period coincided with the coldest part of the Little Ice Age. Maunder was made a fellow of the Royal Astronomical Society in 1873. He died at Greenwich on March 21, 1928.

Maury, Matthew Fontaine (1806–73) American naval officer who became the most distinguished oceanographer and meteorologist of his generation; born on January 14, 1806, in Spotsylvania County, Virginia, and spent his youth in Tennessee. In 1825, when he was 18 years old, Maury joined the U.S. Navy as a midshipman. In 1839 he was injured in a stagecoach accident and rendered permanently lame. No longer fit for active service, in 1841 he was appointed superintendent of the Depot of Charts and Instruments. He remained in this post until 1861. By the time he left, he had transformed the department into the United States Naval Observatory and the Hydrographic Office. Maury issued captains with specially prepared logbooks in which they were asked to record their observations of winds and currents. He charted the course of the Gulf Stream and in 1850 he charted the depths of the North Atlantic Ocean, in order to facilitate laying the transatlantic cable. In 1853 he helped organize a conference on oceanography and meteorology in Brussels, which he attended as the United States representative, and in 1855 he published *Physical Geography of the Sea,* which was the first textbook in oceanography. During the Civil War, Maury served in the Confederate States navy, and after the war he spent some time in exile. He returned home in 1868 to take up the post of professor of meteorology at the Virginia Military Institute. He settled in Lexington, where he died on February 1, 1873. His body was buried temporarily in Lexington and then moved to Hollywood Cemetery, Richmond, where it remains. There is a Maury Hall at the Naval Academy in Annapolis, and in 1930 Maury was elected to the Hall of Fame for Great Americans.

Mie, Gustav (1868–1957) German physicist who discovered the way light is scattered by small particles; born on September 29, 1868, in Rostock. He was educated at the gymnasium (high school) in Rostock and studied mathematics and geology at the Universities of Rostock and Heidelberg. He obtained his doctorate in 1891 at Heidelberg. He worked briefly as a teacher in a private school, then from 1892 until 1902 he was an assistant in the Physics Institute at the Technical University of Karlsruhe. In 1902 he was appointed a special professor (Extraordinariat) at the University of Greifswald, where, in 1908, he wrote his paper on light scattering. He became a professor at the University of Halle in 1918 and in 1924 he became a professor at the University of Freiburg. Mie died at Freiburg im Breisgau on February 13, 1957.

Milankovich, Milutin (1879–1958) This Serbian mathematician and climatologist discovered the link between cyclical changes in the Earth's orbit and rotation; he was born on May 28, 1879, at Dali, near Osijek, in what was then Austria–Hungary. He studied in Vienna, graduating in 1902, and in 1904 took up a post at the University of Belgrade, where he remained for the rest of his career. Milankovich studied the astronomical processes that alter the amount of solar radiation the Earth receives and therefore affect the climate. He identified three relevant cyclical changes and calculated their effects over hundreds of thousands of years. In 1920 he published his results and elaborated on them in subsequent years. The cycles he identified are now known as the Milankovich cycles.

Molina, Mario José (b. 1943) A Mexican atmospheric chemist who helped identify the threat to the ozone layer from chlorofluorocarbon (CFC) compounds, Mario Molina was born in Mexico City on March 19, 1943. He was educated in Switzerland and at the Universidad Nacional Autonoma de Mexico, in Mexico City, graduating in 1965 with a degree in chemical engineering. He obtained a postgraduate degree from the University of Freiburg, Germany in 1967 and his Ph.D. from the University of California at Berkeley in 1972, where F. Sherwood Rowland was one of his supervisors. He held teaching and research posts at the Universidad Nacional Autonoma de Mexico, the University of California at Irvine, and from 1982 to 1989 at the Jet Propulsion Laboratory of the California Institute of Technology, in Pasadena. In 1989 he moved to the Department of Earth, Atmospheric and Planetary Sciences and Department of Chemistry at the Massachusetts Institute of Technology (M.I.T.). He was named M.I.T. Institute Professor in 1997. The first warnings about what might happen to the ozone layer appeared in 1970. This attracted scientific attention and in 1974 Molina and Rowland published a paper describing the results of their studies of the chemistry of CFCs. For this research Mario Molina, Paul Crutzen, and F. Sherwood Rowland were joint winners of the 1995 Nobel Prize for chemistry. More recently, Professor Molina has studied the pollution of the troposphere. In particular, he is keen to find ways to reduce pollution levels in large cities that suffer from traffic congestion, such as his native Mexico City.

Morse, Samuel Finley Breese (1791–1872) American artist and inventor responsible for building the first telegraph line and devising a code by which telegraph messages could be sent simply, efficiently, and reliably; born on April 27, 1791, at Charlestown, Massachusetts. He studied art at Yale University, graduating in 1810, and then traveled to England to study historical painting, staying there from 1811 until 1815. In 1832 Morse became interested in electricity and magnetism, and in the possibility of constructing a telegraph. By about 1835, with considerable help from Joseph Henry, Morse had made a telegraph that worked. He persuaded Congress to appropriate the $30,000 it would cost to build a telegraph line between Baltimore and Washington. He finally succeeded in 1843 and the line was built in 1844. Morse used his own code, which he had developed by 1838, to transmit the first message: "What hath God wrought?" Within a few years telegraphy was being used to transmit meteorological data, a development that led directly to the first weather reports and forecasts. Morse died in New York on April 2, 1872. In 1900, when it first opened, Samuel Morse was made a charter member of the Hall of Fame for Great Americans.

Neumann, John von (originally Janos) (1903–57) This Hungarian-born American mathematician and physicist, who devised a method for numerical weather forecasting, was born in Budapest on December 28, 1903. He was educated privately until he was 14, when he entered the gymnasium (high school) but continued to receive extra instruction in mathematics. He studied at the Universities of Berlin (1921–23) and Zürich (1923–25), graduating from Zürich in chemical engineering. He received a doctorate in mathematics from the University of

Budapest in 1926 and then continued his studies at the University of Göttingen. From 1927 until 1929 von Neumann worked as an unpaid lecturer at the University of Berlin, then moved to the University of Hamburg. He migrated to the United States in 1930 and was appointed a visiting professor at Princeton University. In 1931 he became a full professor of mathematics at the newly formed Institute of Advanced Studies at Princeton. Von Neumann later held a number of advisory posts for the federal government, but he remained at the institute for the rest of his life. In the 1940s von Neumann became interested in meteorology and devised a technique for analyzing the stability of the methods used in numerical forecasting. He established the Meteorology Project at Princeton in 1946 and in April 1950 the group led by von Neumann made the first accurate numerical forecast using the ENIAC computer. In 1952 he designed and supervised the construction of the first computer that was able to use a flexible stored program. This work influenced the design of all the programmable computers that followed, including the computers used in weather forecasting. Von Neumann received many honors and awards, including the Medal of Freedom, the Albert Einstein Award, and the Enrico Fermi Award, all of which were presented to him in 1956. He was already in poor health and he died from cancer in Washington, D.C., on February 8, 1957.

Oeschger, Hans (1927–98) A Swiss physicist whose research revealed ways of reconstructing climates of the remote past, Oeschger trained initially as a nuclear physicist in Zurich and obtained his Ph.D. at the University of Bern in 1955. In collaboration with F. G. Houtermans, in 1955 he built a device to measure very low levels of radioactivity. Known as the Oeschger counter, this was used to date Pacific deep water. In 1962 Oeschger began studying glacier ice. This led to his examinations of antarctic and Greenland ice cores. He used these to trace climate changes over the last 150,000 years. With colleagues Chester C. Langway of the United States and Willi Dansgaard of Denmark he discovered 24 abrupt climate changes, now called Dansgaard–Oeschger events. He also observed changes in the concentration of atmospheric carbon dioxide and its rapid rise in the last 200 years. He was concerned that this might lead to an enhanced greenhouse effect. He was one of the lead authors of the First Report of the Intergovernmental Panel on Climate Change. In 1963 Oeschger founded the Division of Climate and Environmental Physics at the University of Bern and remained its director until he retired in 1992. He died in Bern after a long illness on December 25, 1998.

Pascal, Blaise (1623–62) French mathematician, physicist, and theologian who discovered that the atmosphere has an upper limit and that atmospheric pressure decreases with altitude; born on June 19, 1623, at Clermont-Ferrand, in the Auvergne region. His father, a mathematician and government official, supervised his education. He was a prodigy, who wrote an essay that made René Descartes envious before he was 17, and between 1642 and 1644 he designed and made a calculating machine, based on cogwheels, to help his father. This machine was an ancestor of the modern cash register. Pascal made and sold about 50 of them; several still exist. Learning of the work of Torricelli, Pascal reasoned that if air has weight, the pressure measured by a barometer must decrease with height, because the greater the height of the instrument above the surface the smaller must be the mass of air weighing down from above. He tested this in 1646 at the Puy-de-Dôme, an extinct volcano, 4,806 feet (1,465 m) high. His brother-in-law, Florin Périer, carried a barometer up the mountain and found it showed a steady decrease in pressure with increasing height. A similar barometer at the foot of the mountain showed no change. The experiment confirmed Torricelli's discovery that air has weight. This achievement is recognized in the name of the derived SI unit of pressure or stress, the "pascal." Pascal died in Paris on August 19, 1662, probably from meningitis associated with stomach cancer.

Priestley, Joseph (1733–1804) The English chemist who is usually credited with the discovery of oxygen was born on March 13, 1733, at Fieldhead, Yorkshire. His mother died when he was seven and he went to live with an aunt, who was a Calvinist. Priestley was educated at the Dissenting Academy at Daventry and in 1758 he became a Presbyterian minister. In 1761 he became a language teacher at Warrington Academy. On a visit to London in 1766 he met Benjamin Franklin, who aroused in him an interest in science. From that point he combined his duties as a minister with scientific research. In 1772 he discovered "nitrous air" (nitric oxide, NO) and reduced it to dinitrogen oxide (N_2O), and in the same year he isolated ammonia gas. In 1774 he produced sulfur dioxide (SO_2) by heating concentrated sulfuric acid first with mercury and then with copper. He had found, in 1772, that plants release a gas necessary to animals. In 1774 he found the same gas was released when he heated mercurous oxide (HgO) or red lead (Pb_3O_4). A mouse thrived in it, combustible materials burned brightly in it, and when mixed with NO its volume decreased and a red gas (nitrogen dioxide, NO_2) was produced. Priestley concluded he had discovered air from which the fiery principle phlogiston had been removed (so it readily accepted phlogiston from burning substances) and he called it "dephlogisticated air." In fact Karl Wilhelm Scheele had discovered the gas in 1772, but did not publish the fact. In 1774 Priestley visited Paris, where he met Antoine Lavoisier (1743–94) and told him about dephlogisticated air; it was Lavoisier who called it oxygen. Priestley joined the Lunar Society, a group of eminent scientists based in Birmingham. He continued to voice his dissenting views and his support for the French Revolution. This made him increasingly unpopular with the authorities and in 1794 he immigrated to Northumberland, Pennsylvania. He refused to accept a professorship at the University of Pennsylvania and died at Northumberland on February 6, 1804.

Ramsay, William (1852–1916) The Scottish chemist who received the 1904 Nobel Prize for chemistry for his discoveries of rare gases was born in Glasgow on October 2, 1852. In 1866, when he was only 14, Ramsay entered the University of Glasgow to study arts, and in 1868 he went to work in the laboratory of the Glasgow City Analyst, where he learned chemistry. After a spell in Germany, at the University of Tübingen, where he gained his Ph.D. in chemistry in 1873, Ramsay returned to Glasgow to teach at Anderson's College. In 1880 he became professor of chemistry at University College of Bristol (later Bristol University) and was appointed principal of the college in 1881. In 1887 he became professor of chemistry at

University College, London, a post in which he remained until he retired in 1912. He was knighted in 1902. While at University College, Ramsay discovered argon in 1894, and neon, krypton, and xenon in 1898. He died at High Wycombe, Buckinghamshire, on July 23, 1916.

Rayleigh, Lord (1842–1919) The English physicist who explained, in 1871, why the sky is blue was born as John William Strutt at Terling Place, Langford Grove, Essex, on November 12, 1842. He was taught by a private tutor until he entered Trinity College, Cambridge in 1861, graduating in mathematics in 1865. He then visited the United States, returning to England in 1868 and establishing a private laboratory at his home, Terling Place. On the death of his father in 1873 he became the third baron Rayleigh. Lord Rayleigh, the name by which he is usually known, was elected a fellow of the Royal Society in 1873 and in 1879 became Cavendish Professor of Experimental Physics at Cambridge. After 1884 he spent most of his time working in his private laboratory at home. He was professor of natural philosophy at the Royal Institution, London, from 1887 until 1905. Rayleigh was secretary to the Royal Society from 1885 until 1896 and its president from 1905 until 1908. From 1908 until his death he was chancellor of Cambridge University. Rayleigh spent much of his life studying the properties of light and sound waves, waves in water, and earthquake waves, but his interests were much wider. Measuring the densities of different gases, he found that nitrogen in air was always 0.5% denser than nitrogen from any other source. Unable to account for this, he asked for suggestions. William Ramsay replied and discovered the nitrogen was mixed with argon. Lord Rayleigh won the 1904 Nobel Prize for physics. He died at Terling Place on June 30, 1919.

Renaldini, Carlo (1615–98) An Italian physicist from Ancona, Renaldini was professor of philosophy at the University of Pisa and a member of the Accademia del Cimento (Academy of Experiments) in Florence. In 1694 he proposed a method for calibrating thermometers that he believed would overcome the difficulty caused by the uneven rate at which a liquid expands as it is heated. Renaldini suggested that the points between the two fiducial points could be determined by mixing boiling and ice-cold water in varying proportions. Equal weights of water at 32°F (0°C) and 212°F (100°C) would reveal the halfway point, 20 parts of boiling water mixed with 80 parts of freezing water would indicate 68°F (20°C), and so on. The method proved difficult to put into practice and did not overcome the difficulty of depending on the behavior of a particular liquid.

Reynolds, Osborne (1842–1912) English physicist and engineer whose most important work concerned the flow of water through channels, including wave and tidal movements in rivers and estuaries; born on August 23, 1842, in Belfast, Northern Ireland, but the family moved to Dedham, in Suffolk, England. Osborne was educated at Dedham Grammar School and Queens' College, Cambridge, from where he graduated in mathematics in 1867 and was immediately awarded a fellowship. He worked for a civil engineering company in London, but left in 1868 to become professor of engineering at Owens College, Manchester (now the University of Manchester). Reynolds was elected a fellow of the Royal Society in 1877 and received the Society's Royal Medal in 1888. This was followed by many more medals and honorary degrees. His discoveries led to his formulation of what is now known as the Reynolds number. The Reynolds number is widely used by meteorologists and atmospheric physicists in calculations of air turbulence. Reynolds retired in 1905 and died at Watchet, Somerset, on February 21, 1912.

Richardson, Lewis Fry (1881–1953) The English mathematician and meteorologist who was the first to attempt to forecast weather mathematically was born at Newcastle-upon-Tyne on October 11, 1881. He was educated at a school in York and then studied natural science (physics, mathematics, chemistry, biology, and zoology) at King's College, University of Cambridge. In 1927 he received a doctorate in mathematical psychology at the University of London. (Mathematical psychology is the name given to any theoretical work using mathematical methods, formal logic, or computer simulation.) From 1903 to 1904 he worked for the National Physical Laboratory, and in 1913 he went to work at the Meteorological Office. During his lifetime he was best known for his studies of atmospheric turbulence.

This led to his proposal of what came to be called the Richardson number for predicting whether turbulence would increase or decrease. In 1926 Richardson was elected a fellow of the Royal Society. He described his system for numerical forecasting in his book *Weather Prediction by Numerical Process,* published in 1922. He was ahead of his time. The data needed to construct such forecasts were not available then, though they are now, and there were no calculators, far less supercomputers, to perform the many calculations. Richardson's book was republished in 1965 and a modified version of his method is now used for making large-scale, long-range weather forecasts. Richardson remained a committed Quaker throughout his life. During the First World War, he served among fellow Quakers in the Friends Ambulance Unit. He was sent to France, where he tried to practice the artificial language Esperanto, another interest of his, on German prisoners of war. After the war he returned to the Meteorological Office. A pacifist, in 1920 Richardson resigned from the Meteorological Office when it was absorbed into the Air Ministry. He became head of the Physics Department at Westminster Training College, and in 1929 he was appointed principal of Paisley Technical College (now the University of Paisley) in Scotland. He remained at Paisley until his retirement in 1940. He died at Kilmun, Argyll, Scotland, on September 30, 1953.

Rossby, Carl-Gustav (1898–1957) Swedish-born American meteorologist who discovered the long waves in air movements in the upper atmosphere, now named for him; born in Stockholm, Sweden, on December 28, 1898. He was educated at the University of Stockholm and in 1918, after graduating with a degree in theoretical mechanics, he moved to the Bergen Geophysical Institute. Rossby returned to Stockholm in 1922 to join the Swedish Meteorological Hydrologic Service. In 1926 he visited the United States with a scholarship from the Scandinavian–American Foundation and joined the staff of the Weather Bureau in Washington, D.C. In 1927 he moved to California, and in 1928 he was made the country's first professor of meteorology, at the Massachusetts Institute of Technology. In 1939 he was appointed assistant chief of research and education at the Weather Bureau, but left in 1940 to become chairman of the Institute of Meteorology at the University of Chicago. It was soon after his arrival in Chicago that Rossby developed his theory describing the long atmospheric waves that now bear his name. In 1947 Rossby was invited to establish a department of meteorology at the University of Stockholm. He died in Stockholm on August 19, 1957.

Rowland, Frank Sherwood (b. 1927) American chemist who was one of the three scientists to propose that stratospheric ozone might be depleted by reactions with chlorofluorocarbon compounds (CFCs); born on June 28, 1927, in Delaware, Ohio. He was educated at schools in Delaware and in 1943 enrolled at the Ohio Wesleyan University, studying chemistry, physics, and mathematics. Rowland enlisted in the navy in June 1945, before completing his course, and resumed his studies 14 months later, graduating in 1948. He obtained his Ph.D. in 1952 from the Department of Chemistry at the University of Chicago. From September 1952 until 1956 Rowland was an instructor in the Chemistry Department at Princeton University. He spent three summers, from 1953 to 1955, working on the use of tracer chemicals in the Chemistry Department of the Brookhaven National Laboratory. His link with Brookhaven continued until 1994. He was an assistant professor at the University of Kansas from 1956 until 1964, when he became professor of chemistry at the University of California–Irvine, where he is now Donald Bren Research Professor of Chemistry and Earth System Science. In the early 1970s he became interested in stratospheric ozone. Paul Crutzen had already suggested the possibility that ozone could be depleted. Mario Molina joined Rowland, and in 1974 Rowland and Molina published the results of their research. They had calculated that the breakdown of CFC molecules would release chlorine in a form that would destroy ozone molecules. Their paper stimulated a federal investigation of the situation, and in 1978 the use of CFCs as propellants in spray cans was banned in the United States. The phasing out of the use of CFCs was agreed upon internationally in 1987 under the terms of the Montreal Protocol on Substances That Deplete the Ozone Layer. It was for this work that Rowland, Molina, and Crutzen shared the 1995 Nobel Prize for chemistry.

Rutherford, Daniel (1749–1819) The Scottish chemist who discovered nitrogen was born in Edinburgh on November 3, 1749, and studied medicine at Edinburgh University. In 1772, for his final thesis, Joseph Black, one of his teachers, set him the task of examining air that would not support combustion. Rutherford kept a mouse in an airtight container until it died, then burned a candle and some phosphorus in the same air until they would no longer burn. He passed the air through strong alkali to remove the carbon dioxide and found that the remaining air was still noxious: it would not support combustion and a mouse could not live in it. Believing the air had accepted all the phlogiston it could hold, Rutherford called it "phlogisticated air." Joseph Priestley and Karl Wilhelm Scheele also discovered the gas at about the same time. It was given the name nitrogen in 1790 by the French chemist Jean Antoine Chaptal (1756–1832). In 1786 Rutherford was appointed professor of botany at Edinburgh University. He died in Edinburgh on November 15, 1819.

Saussure, Horace Bénédict de (1740–99) The Swiss physicist who discovered why air temperature decreases with height and who invented the hair hygrometer was born at Conches, near Geneva, on February 17, 1740. He was educated at the public school in Geneva and the Geneva Academy, from where he graduated in 1759 and became professor of philosophy in 1762. He was elected a fellow of the Royal Society in 1772 and in the same year he founded the Society for the Advancement of the Arts in Geneva. He is credited with having constructed the first solar collector, in 1767. When he took this to the top of Mont Cramont he found the outside temperature was 43°F (6°C) but that inside the box was 190°F (88°C). Then he repeated the experiment 4,852 feet (1,480 m) lower down, on the Plains of Cournier. The air temperature there was 77°F (25°C), but the temperature inside the box was almost the same. De Saussure concluded that the Sun shines just as warmly in the mountains as it does on the plains, but that the more transparent mountain air is unable to trap and hold so much warmth. In 1783 de Saussure invented the hair hygrometer, based on his observation that human hairs increase in length as the humidity rises and grow shorter as the air becomes drier. De Saussure resigned from the Geneva

Academy in 1787 and moved to the south of France, where he could live at sea level and collect measurements of atmospheric pressure to compare with those he had taken in the Alps. His health had begun to deteriorate in 1772 and by 1794 he was a sick man. He was also experiencing financial difficulties and was compelled to return with his family to the country house at Conches where he had been born. He died there on the morning of January 22, 1799.

Schaefer, Vincent Joseph (1906–93) American physicist who, with his colleague Bernard Vonnegut, discovered cloud seeding; born at Schenectady, New York, on July 4, 1906. After leaving school he worked for a time in the machine shop at the General Electric Corporation (G.E.C.) in Schenectady. He then studied at Union College, New York, and the Davey Institute of Tree Surgery, from where he graduated in 1928 and became a tree surgeon. Unable to earn an adequate salary at this profession, in 1933 Schaefer returned to G.E.C., where Irving Langmuir (1881–1957) recruited him as a research assistant. Schaefer remained at G.E.C. until 1954. While studying the problem of aircraft icing, Schaefer and Vonnegut studied the formation of ice and snow using a refrigerated box. During a spell of unusually hot weather in July 1946 it was difficult to maintain the temperature inside the box, and on July 13 Schaefer dropped some dry ice (solid carbon dioxide) into the box to chill the air. The moment the dry ice entered the air in the box there was a miniature snowstorm, suggesting a way to make precipitation fall from a cloud that otherwise would not have released it. This discovery led to the development of other techniques for cloud seeding. Schaefer received the degree of doctor of science in 1948 from the University of Notre Dame and in 1959 he joined the faculty of the State University of New York at Albany, where he founded the Atmospheric Sciences Research Center. From 1964 until 1976 he was professor of atmospheric science at the State University of New York. He was appointed a fellow of the American Academy of Arts and Sciences in 1957 and in 1976 he received a special citation from the American Meteorological Society. Schaefer died at Schenectady on July 25, 1993.

Scheele, Karl or **Carl Wilhelm** (1742–86) Swedish chemist who discovered the elements oxygen, chlorine, and nitrogen, as well as a long list of compounds; born in Stralsund, Pomerania, on December 9, 1742. Pomerania was then a part of Sweden, but Scheele spoke and wrote in German. He received little education until he was 14, when he was apprenticed to an apothecary in Göteborg. He was a keen observer, who read and experimented to teach himself chemistry and progressed to increasingly prestigious positions as an apothecary in Malmö (1765), Stockholm (1768), and Uppsala (1770). Other Swedish scientists, including Johann Gahn (1745–1818) and Torbern Bergman (1735–84), recognized his talent. In 1775 Scheele was elected to the Swedish Royal Academy of Sciences and in the same year he moved to Köping, in Västmanland, where he remained for the rest of his life, working as an apothecary. He refused offers of academic appointments in England and Germany and an invitation to become court chemist to Frederick II of Prussia. Not one of his long list of discoveries was fully attributed to him. In some cases other scientists preceded him and in others he failed to complete the research and others claimed the credit; when he discovered chlorine he thought it was an oxygen compound. Nevertheless he was possibly the greatest chemist of the 18th century. His health began to fail in middle age, due partly to overwork and perhaps to his habit of tasting the chemicals he worked with (one of these was hydrogen cyanide, which he discovered). He married while on his deathbed and died at Köping on May 21, 1786, aged 43.

Schönbein, Christian Friedrich (1799–1868) The German-Swiss chemist who discovered ozone was born at Metzingen, Württemberg, on October 18, 1799. He was educated at the Universities of Tübingen and Erlangen and joined the faculty of the University of Basel in 1828. He became a professor in 1835. It was in 1840 that he discovered that the smell people often noticed near electrical equipment could also be produced by the electrolysis of water or oxidizing phosphorus. Schönbein called the gas ozone, from *ozon,* the Greek word for smell. Schönbein also invented guncotton. He died at Sauersberg, Baden, on August 29, 1868.

Schwabe, Heinrich Samuel (1789–1875) The German astronomer who discovered the sunspot cycle was born at Dessau, Anhalt, on October 25, 1789. He was educated in Berlin and became a pharmacist, working in a pharmacy business owned by his mother. He became interested in astronomy while at university and chose an astronomical research topic he could pursue during the day, while still working in the pharmacy. He thought he might discover a new planet inside the orbit of Mercury, detecting it as it crossed in front of the Sun. In 1825 he began observing the Sun with a small telescope and saw sunspots. He became absorbed with these, abandoned the search for a planet, and made daily counts of the sunspots for the rest of his life. He sold the pharmacy in 1829 and became a full-time astronomer. In 1831 he drew a picture of Jupiter that showed the Great Red Spot for the first time. By the end of his life he had published 109 scientific papers and his data filled 31 volumes (presented to the Royal Astronomical Society after his death). In 1857 he received the gold medal of the Royal Astronomical Society, and he became a fellow of the Royal Society in 1868. In 1843, while recognizing the limitations of his equipment, he was able to announce that the number of sunspots waxed and waned over a 10-year cycle (in fact it is 11 years). This finding was ignored until Humboldt mentioned it in 1851, in his book *Kosmos*. Schwabe died at Dessau on April 11, 1875.

Shaw, William Napier (1854–1945) An English meteorologist who did much to establish meteorology as a scientific discipline, Shaw was born in Birmingham on March 4, 1854. He was educated at school in Birmingham and at Emmanuel College, University of Cambridge, from where he graduated in 1876. In 1877 he was elected a fellow of the college and appointed as a lecturer in experimental physics at the Cavendish Laboratory (part of the university). He was appointed assistant director of the Cavendish Laboratory in 1898, but resigned in 1900 to take up an appointment as secretary of the Meteorological Council. He was director of the Meteorological Office from 1905 until his retirement in 1920 and also reader in meteorology at the Royal College of Science of the Imperial College of Science and Technology from 1907 until 1920, and from 1920 until 1924 its first professor of meteorology. Shaw introduced the millibar, in 1909; it was adopted internationally in 1929. Some time about 1915, Shaw devised the tephigram. He pioneered the use of instruments carried beneath kites and balloons to study the upper atmosphere and wrote several books on weather forecasting, as well as one, *The Smoke Problem of Great Cities* (1925), on air pollution. He received many honors, including the 1910 Symons Gold Medal of the Royal Meteorological Society. He was knighted in 1915. Shaw died in London on March 23, 1945.

Simpson, George Clark (1878–1965) A British meteorologist who studied atmospheric electricity and the effect of radiation on polar ice and who assigned wind speeds to the Beaufort wind scale, Simpson was born in Derby on September 2, 1878. He attended school in Derby, leaving in 1894 to work in his father's business, but his reading of popular books about science aroused his interest and he began attending evening classes. After private tutoring he entered Owens College, Manchester, graduating in 1900. He continued his studies at Göttingen University, in Germany, then visited Lapland to study atmospheric electricity. On his return to his college, which by then had become the University of Manchester, he was appointed to head a newly formed meteorology department. He was the first lecturer in meteorology at any British university. In 1905 he was appointed assistant director of the Meteorological Office and was director from 1920 until he retired in 1938. He traveled as meteorologist on Robert Scott's last expedition to Antarctica in 1910 and worked in the Middle East and Egypt between 1916 and 1920. On the outbreak of war, in 1939, he returned to work, and retired for a second time in 1947. He was awarded the Symons Gold Medal of the Royal Meteorological Society in 1930. In 1935 he was knighted. Simpson assigned wind speeds to the Beaufort scale based on measurements from an anemometer 36 ft. (11 m) above the surface (the international standard now uses measurements at 20 ft. [6 m] elevation). He also studied the effect of increasing solar radiation on the polar ice caps, concluding that this initially increases precipitation, causing the ice to advance, then further intensification causes the ice to retreat. As solar radiation decreases, more precipitation falls as snow, causing a second advance of the ice. Simpson died in Bristol on January 1, 1965.

Spörer, Gustav (1822–95) The German solar astronomer who identified a period of low sunspot activity named for him was born in Berlin on October 23, 1822. He studied mathematics and astronomy at the University of Berlin, then worked as a schoolteacher. He began his solar observations in 1858 and quickly established his reputation. In 1874 he joined the Potsdam Astrophysical Laboratory, becoming its chief observer in 1882. By 1887 his studies of records of sunspot activity had convinced him there had been a period during the 17th century when there were very few sunspots. He also found a similar lack of sunspots between about 1400 and 1510, coinciding with a period of severe winters. This is the Spörer minimum. Spörer retired in 1894 and died at Potsdam on July 7, 1895.

Stevin, Simon (1548–1620) The Flemish mathematician who studied the pressure exerted by the weight of liquids was born in Bruges (now in Belgium) in 1548. He is often known as Stevinus, which is the latinized version of his name. He worked as a clerk in Antwerp and then entered the service of the Dutch government, becoming director of the department of roads and waterways and later quartermaster-general to the Dutch army. In 1586 he published *Statics and Hydrostatics,* a book in which he reported that the pressure a liquid exerts on the surface beneath it depends only on the area of that surface and the height to which the liquid extends above the surface. Contrary to what seemed obvious to most people at the time, it has nothing whatever to do with the shape of the vessel holding the liquid. Stevin also established the use of decimal notation in mathematics and maintained that decimal weights, measures, and coinage would eventually be introduced. He died in 1620 either at The Hague or Leiden, two cities 10 miles (16 km) from each other in the Netherlands.

Stewart, Balfour (1828–87) Scottish physicist who studied the radiation and absorption of heat and the properties of blackbodies; born in Edinburgh on November 1, 1828. He was educated at the universities of Dundee and Edinburgh and then joined the staff of the Kew Observatory, near London, becoming its director in 1859. In 1870 he joined the faculty of

Owens College, Manchester (later the University of Manchester). Independently of Gustav Robert Kirchhoff, Stewart found that if the temperature of a body remains constant the amount of radiation it absorbs is equal to the amount it emits. This was the concept of the blackbody. Stewart died near Drogheda, Ireland, on December 19, 1887.

Teisserenc de Bort, Léon Philippe (1855–1913) The French meteorologist who discovered the stratosphere was born in Paris on November 5, 1855. In 1880 he went to work in the meteorological department of the Central Bureau of Meteorology, in Paris, becoming chief meteorologist in 1892. He resigned in 1896 in order to establish a private meteorological observatory at Trappes, near Versailles. Teisserenc de Bort used it primarily to study clouds and the upper air. He was one of the pioneers in the use of balloons to take soundings of the upper atmosphere. These revealed that the air temperature decreased with height up to about 7 miles (11 km). Above this height the temperature remained constant as far as his balloons could reach. In 1900 he proposed that the atmosphere comprises two layers. He called the lower layer the troposphere and its upper boundary the tropopause. Teisserenc de Bort believed that because the temperature above the tropopause remains constant, there is no mechanism such as convection to make the air move vertically. The air in this region might separate into its constituent gases, which would form layers with the heaviest gases at the bottom and the lightest at the top. This stratified arrangement led him to name the isothermal layer the stratosphere. Although the stratosphere is not layered in the way he imagined, the name has survived. Teisserenc de Bort also discovered that the tropopause is much higher over the Tropics than it is over Europe and that European weather is strongly influenced by the distribution of pressure between certain places, especially Iceland and the Azores. He died at Cannes on January 2, 1913.

Thornthwaite, Charles Warren (1899–1963) The American climatologist who devised the system of climate classification bearing his name was born on March 7, 1899, near Pinconning, Michigan. He graduated from Central Michigan Normal School in 1922 as a science teacher and received his doctorate from the University of California–Berkeley, in 1929. Thornthwaite held faculty positions at the University of Oklahoma (1927–34), the University of Maryland (1940–46), and Johns Hopkins University (1946–55). He headed the Division of Climatic and Physiographic Research of the U.S. Soil Conservation Service from 1935 to 1942. From 1946 until his death, he was the director of the Laboratory of Climatology, at Seabrook, New Jersey, and professor of climatology at Drexel Institute of Technology, Philadelphia. The first version of his climate classification appeared in October 1931 and applied only to North America. He published a version covering the world in 1933, with a second version in January 1948 in which he introduced the concept of potential evapotranspiration. C. W. Thornthwaite was one of the most eminent climatologists of his generation, with an international reputation. He was president of the meteorology section of the American Geophysical Union from 1941 to 1944. In 1951 he became president of the Commission on Climatology of the World Meteorological Organization, a post he held until his death. He died from cancer on June 11, 1963.

Torricelli, Evangelista (1608–47) The Italian physicist and mathematician who discovered the principal of the barometer was born at Faenza, near Ravenna, on October 15, 1608. In 1627 he went to study science and mathematics at the Sapienza College, Rome. Having read some of the works of Galileo, in 1641 he wrote a treatise developing some of Galileo's ideas on mechanics. Galileo was impressed and invited him to Florence. Torricelli became his assistant, but Galileo died three months later. Galileo was puzzled by the fact that a piston pump cannot lift water more than 33 feet (10 m) and he asked Torricelli to investigate. Torricelli thought that if the air possessed weight the pressure it exerted outside the cylinder would cause water to rise inside the cylinder as the piston was raised, but perhaps the pressure was sufficient to raise the water by only 33 feet. Torricelli tested this idea in 1643, using mercury, which is denser than water. He partly filled a dish with mercury and completely filled a glass tube 4 feet (1.2 m) long and open at one end. When he inverted the tube until it was upright with its open end beneath the surface of the mercury in the dish, not all the mercury flowed out of the tube and into the dish. The level of mercury in the tube was about 30 inches (760 mm) higher than the surface of the mercury in the dish. Because the mercury descended from a full tube, rather than being drawn upward by a piston, Torricelli had proved that the column of mercury was supported by the pressure exerted by the air on the mercury in the dish and that air has weight. Above the mercury, the tube was empty except for a small amount of mercury vapor. Torricelli was the first person to make a vacuum, and a vacuum produced in this way is still known as a Torricellian vacuum. He noticed that the height of the column of mercury varied slightly from day to day and attributed this to variations in the weight of the air. He had invented a device for measuring these small changes: the first barometer. After Galileo's death, Torricelli was appointed mathematician to the grand duke and professor of mathematics in the Florentine Academy. He died in Florence on October 25, 1647.

Travers, Morris William (1872–1961) The English chemist who collaborated with William Ramsay in the discovery of rare gases was born in Kensington, London, on January 24, 1872. He was educated at schools in Ramsgate, Woking, and Tiverton. In 1889 he enrolled at University College, London, graduating in chemistry in 1893. He became a demonstrator working under Ramsay at University College in 1894, and it was in the following years that the two made their discoveries. Travers received his D.Sc. in 1898 and was made an assistant professor. In 1903 he became a full professor. He was appointed director of the Indian Institute of Scientists, in Bangalore, India, in 1906, returning to England at the outbreak of war in 1914 to direct glass production at a factory in London. He became professor of chemistry at Bristol University in 1927 and remained there until he retired in 1937. During World War II he was an adviser and consultant on explosives to the Ministry of Supply. He died at Stroud, Gloucestershire, on August 25, 1961.

Tyndall, John (1820–93) The Irish physicist who was the first to suggest the greenhouse effect was born at Leighlin Bridge, County Carlow, on August 2, 1820. He attended school in Carlow, leaving at 17, and in 1839 he joined the Ordnance

Survey, the British government agency responsible for mapping the country, to train as a surveyor. From 1843 until 1847 he conducted surveys for the companies that were rapidly expanding the railroad network in England. In 1847 Tyndall found a post teaching mathematics, and in October 1848 he and his friend Edward Frankland (1825–99) enrolled at the University of Marburg, Germany. Tyndall studied physics, calculus, and chemistry, then specialized in the study of magnetism and optics. After receiving his doctorate in 1850 he continued his research for an additional year. In 1851, short of funds, Tyndall returned to his post at Queenwood College. He was elected a fellow of the Royal Society in 1852, and in 1853 it was arranged that he should give a lecture at the Royal Institution in London. This was so successful that he was invited to give a second lecture, and then a whole course of lectures. A few months later he was appointed professor of natural philosophy at the Royal Institution. When Michael Faraday (1791–1867) died Tyndall succeeded him as superintendent of the Royal Institution. From 1859 to 1868 Tyndall was also professor of physics at the Royal School of Mines, where he gave a series of popular lectures on science, working in collaboration with a close friend, Thomas Henry Huxley (1825–95). In January 1859 Tyndall began his research into radiant heat. He found that although oxygen, nitrogen, and hydrogen are completely transparent to radiant heat, other gases, especially water vapor, carbon dioxide, and ozone, are relatively opaque. Tyndall said that without water vapor the surface of the Earth would be permanently frozen. Later he speculated about the ways in which changing the concentrations of these gases might affect the climate. This was the first suggestion of what is now known as the greenhouse effect. In 1869, while investigating the way light passes through liquids, Tyndall suggested that atmospheric particles should scatter light and air molecules should scatter blue light more than red light, causing the blue color of the sky. Lord Rayleigh confirmed this in 1871. Tyndall used the scattering effect to measure the pollution of London air, and he was able to show, in 1881, that bacterial spores are present in even the most carefully filtered air. Throughout his adult life Tyndall slept badly and was often unwell. By the 1880s his health was deteriorating. He retired from the Royal Institution in 1887 and went to live at Hindhead, Surrey. His insomnia became steadily worse and he experimented with a variety of drugs to treat it. He died on December 4, 1893, after his wife had accidentally given him an overdose of the sedative chloral hydrate.

Vonnegut, Bernard (1914–97) American physicist who, with Vincent Schaefer, discovered cloud seeding; born at Indianapolis, Indiana, on August 29, 1914. He graduated in 1936 from the Massachusetts Institute of Technology (M.I.T.) and received his Ph.D. from M.I.T. in 1939 for research into aircraft icing. From 1939 until 1941 he worked for the Hartford Empire Company and was a research associate at M.I.T. from 1941 until 1945. In 1945 Vonnegut moved to the laboratories of the General Electric Corporation in Schenectady, New York, where he continued his research into icing. Vonnegut moved to the Arthur D. Little Corporation in 1952, and in 1967 he was appointed Distinguished Research Professor of the State University of New York, a position he held until his death. Following the discovery that dry ice (solid carbon dioxide) was effective at cloud seeding, Vonnegut turned his attention to the search for other materials that might perform the same task and discovered silver iodide crystals were effective. Vonnegut died at Albany, New York, on April 25, 1997.

Wegener, Alfred Lothar (1880–1930) German meteorologist and geologist best known for his theory of continental drift; born in Berlin on November 1, 1880. He was educated at the Universities of Heidelberg, Innsbruck, and Berlin. In 1905 he received a Ph.D. in planetary astronomy from the University of Berlin, but immediately switched to meteorology, taking a job at the Royal Prussian Aeronautical Observatory, near Berlin. He was primarily a meteorologist and studied the formation of raindrops and the circulation of air over polar regions. He used kites and balloons to study the upper atmosphere. In 1906 he joined a two-year Danish expedition to Greenland as the official meteorologist. Wegener studied the polar air, using kites and tethered balloons, and on his return to Germany in 1909 he became a lecturer in meteorology and astronomy at the University of Marburg. His book *Thermodynamik der Atmosphäre* (Thermodynamics of the atmosphere), published in 1911, became a standard textbook throughout Germany. In it he proposed a method of raindrop formation that was confirmed in the 1930s by Tor Bergeron and Walter Findeisen. It is sometimes called the Wegener–Bergeron–Findeisen mechanism. Wegener was also the first person to explain what are now called Wegener arcs, occasionally seen in the arctic. In 1912 Wegener married Else Köppen, the daughter of Wladimir Köppen. Wegener and Köppen collaborated on a book about the history of climate, *The Climates of the Geological Past.* Wegener then returned to Greenland. The four-person 1912–13 expedition crossed the ice cap and was the first to spend the winter on the ice. In 1924 he accepted a post created especially for him and became the professor of meteorology and geophysics at the University of Graz, in Austria. In 1930 he returned to Greenland as the leader of a team of 21 scientists and technicians planning to study the climate over the ice cap. It was during this expedition that Wegener disappeared, some time after November 1, 1930, while returning from a camp on the ice to which he had taken supplies. His body was found on May 12, 1931. He appeared to have suffered a heart attack.

SECTION THREE
CHRONOLOGY

c. 340 BCE ● Greek philosopher Aristotle (384–322 BCE) writes *Meteorologica,* the oldest known work on meteorology and possibly the first. It gives us the word *meteorology.*

140–131 BCE ● Han Ying, in China, writes *Moral Discourses Illustrating the Han Text of the "Book of Songs,"* containing the first known description of the hexagonal structure of snowflakes.

1st century BCE ● The Tower of the Winds is built in Athens. It predicts weather on the basis of wind direction and is probably the first forecasting device.

c. 55 BCE ● Possibly the first person to notice that thunder is always associated with big, solid-looking clouds, Roman poet and philosopher Titus Lucretius Carus (c. 95–55 BCE) proposes that thunder is the sound of great clouds crashing together.

1st century CE ● Greek writer Hero (or Heron) of Alexandria (born about 20 CE) demonstrates that air is a substance.

1555 ● Swedish priest and naturalist Olaus Magnus (1490–1557) publishes a book containing the first European drawings of ice crystals and snowflakes.

1586 ● Flemish mathematician Simon Stevin (1548–1620) shows that the pressure a liquid exerts on a surface depends on the height of the liquid above the surface and the area of the surface on which it presses, but not on the shape of the vessel containing the liquid.

1593 ● Italian physicist and astronomer Galileo Galilei (1564–1642) invents his air thermoscope.

1611 ● German astronomer Johannes Kepler (1571–1630) publishes *A New Year's Gift, or On the Six-cornered Snowflake* in which he describes snowflakes.

1641 ● Italian physicist (and grand duke of Tuscany) Ferdinand II (1610–70) invents a thermometer consisting of a sealed tube containing liquid.

1643 ● Italian physicist and mathematician Evangelista Torricelli (1608–47) invents the mercury barometer.

1646 ● French mathematician, physicist, and theologian Blaise Pascal (1623–62) demonstrates that atmospheric pressure decreases with height.

1654 ● Ferdinand II produces an improved version of his thermometer.

1660 ● English physicist Robert Boyle (1627–91) publishes his discovery of the relationship between the volume occupied by a gas and the pressure under which the gas is held.

1686 ● English astronomer Edmund Halley (1656–1742) proposes the first explanation for the trade winds.

1687 ● French physicist Guillaume Amontons (1663–1705) invents the hygrometer.

1714 ● Polish-Dutch physicist Daniel Fahrenheit (1686–1736) invents the mercury thermometer and the temperature scale bearing his name.

1735 ● English meteorologist George Hadley (1685–1768) proposes a model of the atmospheric circulation to explain the direction from which the trade winds blow.

1738 ● Swiss natural philosopher Daniel Bernoulli (1700–82) demonstrates that when the velocity of a flowing fluid increases its internal pressure decreases.

1742 ● Swedish astronomer and physicist Anders Celsius (1701–44) proposes the temperature scale that will later bear his name.

1752 ● American statesman, physicist, and inventor Benjamin Franklin (1706–90) performs his experiment with a kite, demonstrating that storm clouds carry electric charge and that a lightning stroke is a giant spark.

1761 ● Scottish chemist Joseph Black (1728–99) discovers latent heat.

1783 ● Swiss physicist Horace Bénédict de Saussure (1740–99) invents the hair hygrometer.

1806 ● English Admiral Francis Beaufort (1774–1857) proposes a scale for classifying wind forces.

1820 ● English meteorologist, chemist, and inventor John Daniell (1790–1845) invents the dewpoint hygrometer.

1824 ● John Daniell shows the importance of maintaining a humid atmosphere in hothouses growing tropical plants.

1827 ● French mathematician and physicist Jean-Baptiste Fourier (1768–1830) writes what may be the first account of the greenhouse effect.

1835 ● French physicist Gaspard de Coriolis (1792–1843) explains that bodies moving over the surface of the Earth, but not attached to it, are deflected by inertia acting at right angles to their direction of motion.

1840 ● Swiss-American naturalist Louis Agassiz (1807–73) discovers that glaciers move and once covered a much larger area than they do now.

1842 ● American oceanographer and meteorologist Matthew Maury (1806–73) discovers the shape of storms from data gathered from ships at sea.

● Austrian physicist Christian Doppler (1803–53) discovers that the pitch of a sound rises and light becomes bluer if the source is approaching, and the pitch of a sound falls and light becomes redder if the source is receding, an effect that will later be named after him.

1844 ● The world's first telegraph line opens between Baltimore and Washington.

1846 ● American physicist Joseph Henry (1797–1878) is elected secretary of the Smithsonian Institution; he uses his position to obtain weather reports from all over the United States.

1851 ● The first weather map is exhibited, at the Great Exhibition in London, England.

1855 ● French astronomer Urbain Jean Joseph Leverrier (1811–77) begins supervising the installation of a network to gather meteorological data from observatories across Europe.

1856 ● American climatologist William Ferrel (1817–91) completes the explanation for the direction of the trade winds. He also proposes that in the Northern Hemisphere winds blow counterclockwise around areas of low pressure.

1857 ● Dutch meterologist C. D. H. Buys Ballot (1817–90) proposes the law that later bears his name (although it was discovered earlier by William Ferrel).

1861 ● The Meteorological Department of the Board of Trade issues the first British storm warnings for coastal areas on February 6 and for shipping on July 31.

● Irish physicist John Tyndall (1820–93) shows that certain atmospheric gases absorb heat and therefore that the chemical composition of the atmosphere affects climate.

1863 ● The first network of meteorological stations linked by telegraph to a central point opens in France.

● English scientist and inventor Francis Galton (1822–1911) devises a method for mapping weather systems and coins the term *anticyclone*.

1869 ● The first daily weather bulletins begin to be issued from Cincinnati Observatory on September 1.

1871 ● The first three-day weather forecasts are issued by the Weather Bureau.

1874 ● The International Meteorological Congress is founded.

1875 ● A weather map appears in a newspaper, the *Times* of London, for the first time.

1884 ● American astronomer and physicist S. P. Langley (1834–1906) publishes a paper on the climatic effect of the absorption of heat by atmospheric gases.

1891 ● The U.S. Weather Bureau is founded.

1893 ● English astronomer Edward Maunder (1851–1928) discovers the link between solar activity and the Little Ice Age.

1896 ● Swedish physical chemist Svante Arrhenius (1859–1927) links climatic changes to the atmospheric concentration of carbon dioxide.

      ● The International Meteorological Congress publishes the first edition of the *International Cloud Atlas.*

1902 ● French meteorologist L. P. Teisserenc de Bort (1855–1913) discovers the stratosphere.

      ● Norwegian physicist and meteorologist Vilhelm Bjerknes (1862–1951) publishes one of the first scientific studies of weather forecasting.

1905 ● Swedish oceanographer and physicist V. W. Ekman (1874–1954) discovers that the deflection of winds and ocean currents changes with vertical distance from the surface.

1913 ● French physicist Charles Fabry (1867–1945) discovers the ozone layer.

1918 ● German meteorologist and climatologist W. P. Köppen (1846–1940) publishes a system for classifying climates.

      ● Vilhelm Bjerknes establishes the existence of air masses.

1922 ● English mathematician and meteorologist L. F. Richardson (1881–1953) describes a method for numerical forecasting.

1923 ● English meteorologist Sir Gilbert Thomas Walker (1868–1958) describes the high-level flow of air from west to east close to the equator and the southern oscillation linked to El Niño events.

1930 ● Serbian mathematician and climatologist Milutin Milankovich (1879–1958) proposes a link between variations in the Earth's orbit and rotation and the onset and ending of glacial periods.

1931 ● American farmer and photographer Wilson Bentley (1865–1931) publishes more than 2,000 photographs of snowflakes.

● American climatologist C. W. Thornthwaite (1899–1963) publishes a system for classifying climates.

1940 ● Swedish-American meteorologist Carl-Gustav Rossby (1898–1957) discovers large-wavelength undulations in the westerly winds of the upper atmosphere.

1946 ● American physicist Vincent Schaefer (1906–93) discovers that pellets of dry ice (solid carbon dioxide) can trigger the formation of ice crystals.

1949 ● Radar is used for the first time to obtain meteorological data.

1951 ● An international system for the classification of snowflakes is adopted.

1959 ● The U.S. Weather Bureau begins publishing a temperature-humidity index as an indication of how comfortable the air will feel on a warm day.

1960 ● The first weather satellite, *Tiros 1*, is launched.

1961 ● American meteorologist Edward Norton Lorenz (born 1917) discovers that weather systems behave chaotically; he calls their extreme sensitivity to small variations in initial conditions the "butterfly effect."

1964 ● The *Nimbus 1* weather satellite is launched.

1966 ● The first satellite to be placed in a geostationary orbit is launched on December 6.

1971 ● The Fujita Tornado Intensity Scale is published.

1973 ● Doppler radar is used successfully for the first time to study a tornado.

1974 ● The first *GOES (Geostationary Operational Environmental Satellite)* is launched.

- American chemist F. Sherwood Rowland (b. 1927) and Mexican atmospheric chemist Mario Molina (b. 1943) propose that CFCs might deplete stratospheric ozone.

1985 - Depletion of the ozone layer over Antarctica is discovered by J. C. Farman, B. G. Gardiner, and J. D. Shanklin.

1987 - The Montreal Protocol on Substances That Deplete the Ozone Layer is agreed upon, aiming to reduce and eventually eliminate the release into the atmosphere of all human-made substances that deplete stratospheric ozone.

1988 - The Intergovernmental Panel on Climate Change (IPCC) is founded by the U.N. Environment Program (UNEP) and the World Meteorological Organization (WMO).

1990 - The IPCC publishes its first report on climate change, warning of a temperature rise of 3.24°F (1.8°C) by 2020 and of 4.3–9.2°F (2.4–5.1°C) by 2070.

1992 - The Framework Convention on Climate Change is agreed upon at the United Nations Conference on Environment and Development held in Rio de Janeiro, Brazil; it aims to promote research into climate change and address the causes.

- The IPCC publishes its second report, warning of a temperature rise of 2.7–8.1°F (1.5–4.5°C) by 2100.

1993 - Using powerful computers and advanced climate models, the National Weather Service is able to predict a major storm five days in advance.

1995 - F. Sherwood Rowland, Mario Molina, and Dutch atmospheric chemist Paul Crutzen (born 1933) share the Nobel Prize in chemistry for their discovery that CFCs may deplete the stratospheric ozone layer.

1996 - Technological advances mean five-day forecasts are as accurate as three-day forecasts had been in 1980.

- The IPCC publishes its third report, claiming that global warming has been detected and warning of a temperature rise of 1.8–6.3°F (1–3.5°C) by 2100.

1997 ● Representatives from 160 nations agree on the Kyoto Protocol on Climate Change, aimed at reducing emissions of greenhouse gases.

2001 ● The IPCC publishes its fourth report, warning of a temperature rise of 2.5–10.4°F (1.4–5.8°C) by 2100.

SECTION FOUR
CHARTS & TABLES

Composition of the present atmosphere

Gas	Chemical formula	Abundance
Major constituents		
nitrogen	N_2	78.08%
oxygen	O_2	20.95%
argon	Ar	0.93%
water vapor	H_2O	variable
Minor constituents		
carbon dioxide	CO_2	365 p.p.m.v.
neon	Ne	18 p.p.m.v.
helium	He	5 p.p.m.v.
methane	CH_4	2 p.p.m.v.
krypton	Kr	1 p.p.m.v.
hydrogen	H_2	0.5 p.p.m.v.
nitrous oxide	N_2O	0.3 p.p.m.v.
carbon monoxide	CO	0.05–0.2 p.p.m.v.
xenon	Xe	0.08 p.p.m.v.
ozone	O_3	variable
Trace constituents		
ammonia	NH_3	4 p.p.b.v.
nitrogen dioxide	NO_2	1 p.p.b.v.
sulfur dioxide	SO_2	1 p.p.b.v.
hydrogen sulfide	H_2S	0.05 p.p.b.v.

Atmospheric residence times

Substance	Symbol	Atmospheric residence time
nitrogen	N_2	42 million years
oxygen	O_2	1,000 years
CFC-114	$C_2F_4Cl_2$	300 years
HCFC-23	CHF_3	250 years
CFC-12 (Freon-12)	CF_2Cl_2	100 years
CFC-113	$C_2F_3Cl_3$	85 years
H-1301	CF_3Br	65 years
carbon dioxide	CO_2	55 years
CFC-11 (Freon-11)	$CFCl_3$	45 years
carbon tetrachloride	CCl_4	35 years
HCFC-142b	CH_3CClF_2	19 years
H-1211	CF_2ClBr	16 years
HCFC-22	CHF_2Cl	12 years
methane	CH_4	11 years
HCFC-141b	CH_3CCl_2F	9 years
HCFC-124	C_2HF_4Cl	5.9 years
methyl chloroform	CH_3CCl_3	4.8 years
methyl chloride	CH_3Cl	1.5 years
HCFC-123	$C_2HF_3Cl_2$	1.4 years
water	H_2O	10 days
ammonia	NH_3	7 days
smoke		hours
large particles		minutes

Albedo

Surface	Value
Fresh snow	0.75–0.95
Old snow	0.40–0.70
Cumuliform cloud	0.70–0.90
Stratiform cloud	0.59–0.84
Cirrostratus	0.44–0.50
Sea ice	0.30–0.40
Dry sand	0.35–0.45
Wet sand	0.20–0.30
Desert	0.25–0.30
Meadow	0.10–0.20
Field crops	0.15–0.25
Deciduous forest	0.10–0.20
Coniferous forest	0.05–0.15
Concrete	0.17–0.27
Black road	0.05–0.10

Air density and height

Height (km)	(miles)	Density (kg m^{-3})	(lb. ft.$^{-3}$)
30	18.6	0.02	0.0012
25	15.5	0.04	0.0025
20	12.4	0.09	0.0056
19	11.8	0.10	0.0062
18	11.2	0.12	0.0075
17	10.6	0.14	0.0087
16	9.9	0.17	0.0106
15	9.3	0.20	0.0125
14	8.7	0.23	0.0143
13	8.1	0.27	0.0169
12	7.5	0.31	0.0193
11	6.8	0.37	0.0231
10	6.2	0.41	0.0256
9	5.6	0.47	0.0293
8	5.0	0.53	0.0331
7	4.3	0.59	0.0368
6	3.7	0.66	0.0412
5	3.1	0.74	0.0462
4	2.5	0.82	0.0512
3	1.9	0.91	0.0568
2	1.2	1.01	0.0630
1	0.6	1.11	0.0693
0	0	1.23	0.0768

Typical emissivities

Water	0.92–0.96
Fresh snow	0.82–0.995
Desert	0.90–0.91
Tall, dry grass	0.90
Oak woodland	0.90
Pine forest	0.90
Dry concrete	0.71–0.88
Plowed field	0.90

Specific heat capacities *(c)* of common substances

Substance	Temperature		c	
	°C	°F	J g^{-1} K^{-1}	cal g^{-1} °C^{-1}
fresh water	15	59	4.19	1.00
sea water	17	62.6	3.93	0.94
ice	−21 – −1	−5.8–30.2	2.0–2.1	0.48–0.50
dry air	20	68	1.006	0.2403
basalt	20–100	68–212	0.84–1.00	0.20–0.24
granite	20–100	68–212	0.80–0.84	0.19–0.20
white marble	18	64.4	0.88–0.92	0.21–0.22
quartz	0	32	0.73	0.17
sand	20–100	68–212	0.84	0.20

Distribution of fresh water

Location	Percent of total fresh water
ice caps and glaciers	75
groundwater	22
upper soil	1.75
lakes and inland seas	0.6
rivers	0.003

The boiling temperature of water at a range of pressures

Boiling temperature		Atmospheric pressure
°F	°C	mb
392	200	15,536
320	160	6,176.8
248	120	1,984.9
212	100	1,013.25
140	60	199.33
68	20	23.38
32	0	6.11

Saturation vapor pressure

Temperature °F (°C)	Pressure mb (Pa)
−58 (−50)	0.039 (3.94)
−40 (−40)	0.128 (12.83)
−22 (−30)	0.380 (37.98)
−4 (−20)	1.032 (103.2)
14 (−10)	2.597 (259.7)
32 (0)	6.108 (610.78)
50 (10)	12.272 (1,227.2)
68 (20)	23.373 (2,337.3)
86 (30)	42.430 (4,243.0)
104 (40)	73.777 (7,377.7)

Saturation mixing ratio and temperature (at sea-level pressure)

Temperature °F (°C)	Saturation mixing ratio g kg^{-1}
104 (40)	47
95 (35)	35
86 (30)	26.5
77 (25)	20
68 (20)	14
59 (15)	10
50 (10)	7
41 (5)	5
32 (0)	3.5
14 (−10)	2
−4 (−20)	0.75
−22 (−30)	0.3
−40 (−40)	0.1

Mean snow line

Latitude	Northern Hemisphere (feet)	(meters)	Southern Hemisphere (feet)	(meters)
0–10	15,500	4,727	17,400	5,310
10–20	15,500	4,727	18,400	5,610
20–30	17,400	5,310	16,800	5,125
30–40	14,100	4,300	9,900	3,020
40–50	9,900	3,020	4,900	1,495
50–60	6,600	2,010	2,600	793
60–70	3,300	1,007	0	0
70–80	1,650	503	0	0

Converting snowfall to rainfall equivalent

Snow to water ratios

Temperature °F	°C	Ratio
35	1.7	7:1
29–34	−1.7–1.1	10:1
20–28	−6.7 − −2.2	15:1
10–19	−12.2 − −7.2	20:1
0–9	−17.8 − −12.8	30:1
less than 0	less than −17.8	40:1

Cloud classification

Cloud level	Height of base					
	Polar regions		Temperate latitudes		Tropics	
	'000 feet	'000 meters	'000 feet	'000 meters	'000 feet	'000 meters
High cloud:	10–26	3–8	16–43	5–13	16–59	5–18
Types: CIRRUS, CIRROSTRATUS, CIRROCUMULUS						
Middle cloud:	6.5–13	2–4	6.5–23	2–7	6.5–26	2–8
Types: ALTOCUMULUS, ALTOSTRATUS, NIMBOSTRATUS						
Low cloud:	0–6.5	0–2	0–6.5	0–2	0–6.5	0–2
Types: STRATUS, STRATOCUMULUS, CUMULUS, CUMULONIMBUS						

UV Index

UV category	UVI value	Time to burn (minutes)	Precautions
Minimal	0–2	30–60	Wear a hat
Low	3–4	15–20	Wear a hat; use sunscreen SPF 15+
Moderate	5–6	10–12	Wear a hat; use sunscreen SPF 15+; keep in shade
High	7–9	7–8.5	Wear a hat; use sunscreen SPF 15+; keep in shade; stay indoors between 10 A.M. and 4 P.M.
Very high	10–15	4–6	Stay indoors as much as possible; outdoors wear a hat and use sunscreen SPF 15+

Palmer Drought Severity Index Classification

4.00 or more	Extremely wet
3.00–3.99	Very wet
2.00–2.99	Moderately wet
1.00–1.99	Slightly wet
0.50–0.99	Incipient wet spell
0.49 – –0.49	Near normal
–0.50 – –0.99	Incipient dry spell
–1.00 – –1.99	Mild drought
–2.00 – –2.99	Moderate drought
–3.00 – –3.99	Severe drought
–4.00 or less	Extreme drought

Beaufort Wind Scale

Force	Speed mph (kmh)	Name	Description
0	0.1 (1.6) or less	Calm	Air feels still. Smoke rises vertically.
1	1–3 (1.6–4.8)	Light air	Wind vanes and flags do not move, but rising smoke drifts.
2	4–7 (6.4–11.2)	Light breeze	Drifting smoke indicates the wind direction.
3	8–12 (12.8–19.3)	Gentle breeze	Leaves rustle, small twigs move, and flags made from lightweight material stir gently.
4	13–18 (20.9–28.9)	Moderate breeze	Loose leaves and pieces of paper blow about.
5	19–24 (30.5–38.6)	Fresh breeze	Small trees that are in full leaf sway in the wind.
6	25–31 (40.2–49.8)	Strong breeze	It becomes difficult to use an open umbrella.
7	32–38 (51.4–61.1)	Moderate gale	The wind exerts strong pressure on people walking into it.
8	39–46 (62.7–74)	Fresh gale	Small twigs torn from trees.
9	47–54 (75.6–86.8)	Strong gale	Chimneys are blown down. Slates and tiles are torn from roofs.
10	55–63 (88.4–101.3)	Whole gale	Trees are broken or uprooted.
11	64–75 (102.9–120.6)	Storm	Trees are uprooted and blown some distance. Cars are overturned.
12	more than 75 (120.6)	Hurricane	Devastation is widespread. Buildings are destroyed and many trees are uprooted.

Saffir/Simpson Hurricane Scale

Category	Pressure at center mb in. of mercury cm of mercury	Wind speed mph kmh	Storm surge feet meters	Damage
1	980 28.94 73.5	74–95 119–153	4–5 1.2–1.5	Trees and shrubs lose leaves and twigs. Mobile homes destroyed.
2	965–979 28.5–28.91 72.39–73.43	96–110 154.4–177	6–8 1.8–2.4	Small trees blown down. Exposed mobile homes severely damaged. Chimneys and tiles blown from roofs.
3	945–964 27.91–28.47 70.9–72.31	111–130 178.5–209	9–12 2.7–3.6	Leaves stripped from trees. Large trees blown down. Mobile homes demolished. Small buildings damaged structurally.
4	920–944 27.17–27.88 69.01–70.82	131–155 210.8–249.4	13–18 3.9–5.4	Extensive damage to windows, roofs, and doors. Mobile homes destroyed completely. Flooding to 6 miles (10 km) inland. Severe damage to lower parts of buildings near exposed coasts.
5	920 or lower below 17.17 below 69	more than 155 more than 250	more than 18 more than 5.4	Catastrophic. All buildings severely damaged, small buildings destroyed. Major damage to lower parts of buildings less than 15 feet (4.6 m) above sea level to 0.3 mile (0.5 km) inland.

Fujita tornado intensity scale

Rating	Wind speed		Damage
	mph	kmh	
Weak			
F-0	40–72	64–116	Slight
F-1	73–112	117–180	Moderate
Strong			
F-2	113–157	182–253	Considerable
F-3	158–206	254–331	Severe
Violent			
F-4	207–260	333–418	Devastating
F-5	261–318	420–512	Incredible

Hailstorm intensity scale

Intensity	Damage	Hailstone size
H0	None	5–10 mm (0.2–0.4 in.)
H1	Makes holes in leaves and flower petals	5–20 mm (0.2–0.8 in.)
H2	Strips leaves from plants, damages vegetables	5–30 mm (0.2–1.2 in.)
H3	Breaks glass panes, scrapes paint, marks woodwork, dents trailers, tears tents	11–45 mm (0.4–1.8 in.)
H4	Breaks windows, cracks windscreens, scrapes off paint, kills chickens and small birds	16–60 mm (0.6–2.4 in.)
H5	Breaks some roof tiles and slates, dents cars, strips bark from trees, cuts branches from trees, kills small animals	21–80 mm (0.8–3 in.)
H6	Breaks many roof tiles and slates, cuts through roof shingles and thatch, makes some holes in corrugated iron, breaks wooden window frames	31–100 mm (1.2–3.9 in.)
H7	Shatters many roofs, breaks metal window frames, seriously damages car bodies	46–125 mm (1.8–4.9 in.)
H8	Cracks concrete roofs, destroys other roofs, marks pavements, splits tree trunks, can seriously injure people	61–more than 125 mm (2.4–more than 4.9 in.)
H9	Marks concrete walls, makes holes in walls of wooden houses, fells trees, can kill people	81–more than 125 mm (3.2–more than 4.9 in.)
H10	Destroys wooden houses, seriously damages brick houses, can kill people	101–more than 125 mm (4–more than 4.9 in.)

(H = Hail)

Avalanche classes

(There are five classes. Each class is ten times stronger than the one preceding it.)

Class	Damage	Path width
1	Could knock someone over, but not bury them.	10 m (33 ft.)
2	Could bury, injure, or kill someone.	100 m (330 ft.)
3	Could bury and wreck a car, damage a truck, demolish a small building, break trees.	1,000 m (3,330 ft.)
4	Could wreck a railroad car or big truck, demolish several buildings, . or up to 4 ha (10 acres) of forest.	2,000 m (6,560 ft.)
5	Largest known; could destroy a village or up to 40 ha (100 acres) of forest.	3,000 m (9,800 ft.)

Present weather codes

00   No cloud developing during the past hour
01   Cloud dissolving during the past hour
02   Cloud generally unchanged during the past hour
03   Cloud developing during the past hour
04   Visibility reduced by smoke
05   Haze
06   Dust widespread
07   Dust or sand raised by local wind, but not by dust storms, sandstorms, or whirls (devils)
08   Dust or sand whirls seen in the past hour, but no dust storms or sandstorms
09   Dust storm or sandstorm seen nearby during the past hour
10   Mist
11   Shallow, patchy fog or ice fog
12   Shallow, continuous fog or ice fog
13   Lightning but no thunder
14   Precipitation seen, but not reaching the surface
15   Precipitation seen reaching the surface in the distance
16   Precipitation seen reaching the surface nearby, but not at the station
17   Thunderstorm but no precipitation seen
18   Squalls at the time of observation or during the past hour
19   Funnel cloud seen at the time of observation or during the past hour
20   Precipitation, fog, or thunderstorm during the past hour but not at the time of observation
21   Drizzle (not freezing) or snow grains, but not in showers
22   Rain (not freezing), but not in showers
23   Rain and snow or ice pellets, but not in showers
24   Freezing drizzle or freezing rain
25   Rain showers
26   Showers of rain and snow (British sleet) or snow
27   Showers of hail and rain or hail
28   Fog or ice fog in the past hour

(continues)

Present weather codes *(continued)*

29	Thunderstorm
30–39	Dust storms, sandstorms, drifting snow, or blowing snow
40–49	Fog or ice fog at the time of observation
50	Drizzle (not freezing) that is intermittent and slight at the time of observation
51	Drizzle (not freezing) that is continuous at the time of observation
52	Drizzle (not freezing) that is intermittent and moderate at the time of observation
53	Drizzle (not freezing) that is continuous and moderate at the time of observation
54	Drizzle (not freezing) that is intermittent and heavy at the time of observation
55	Drizzle (not freezing) that is continuous and heavy at the time of observation
56	Slight freezing drizzle
57	Moderate or heavy freezing drizzle
58	Slight drizzle and rain
59	Moderate or heavy drizzle and rain
60–69	The same as 50–59, but with rain instead of drizzle and in 68 and 69 snow instead of rain
70	Snowflakes, intermittent and slight at the time of observation
71–75	The same as 51–55, but with snow instead of drizzle
76	Ice prisms with or without fog
77	Snow grains with or without fog
78	Isolated, star-like snow crystals with or without fog
79	Ice pellets
80	Rain showers, slight
81	Rain showers, moderate or heavy
82	Rain showers, violent
83	Rain and snow showers, slight
84	Rain and snow showers, moderate or heavy
85	Snow showers, slight
86	Snow showers, moderate or heavy
87	Slight showers of snow pellets, encased in ice or not, with or without rain or rain and snow (British sleet) showers
88	Moderate showers of snow pellets, encased in ice or not, with or without rain or rain and snow (British sleet) showers
89	Slight hail showers, without thunder, with or without rain or rain and snow (British sleet)
90	Moderate or heavy hail showers, without thunder, with or without rain or rain and snow (British sleet)
91	Slight rain
92	Moderate or heavy rain
93	Slight snow, or rain and snow (British sleet), or hail
94	Moderate or heavy snow, or rain and snow (British sleet), or hail
95	Slight or moderate storm with rain and/or snow, but no hail
96	Slight or moderate storm with hail
97	Heavy storm with rain and/or snow, but no hail
98	Storm with sandstorm or dust storm
99	Heavy storm with hail

Cyclone names

(If a tropical cyclone has a major impact, the country or countries most affected may ask the World Meteorological Organization to retire its name from the list. This allows the name to be associated unambiguously with a particular storm in historical references and for the purposes of insurance claims and legal actions. Once retired, a name cannot be used for at least 10 years. An alternative name of the same gender and language—English, French, or Spanish—is then substituted for the retired name.)

Atlantic

2003	2004	2005	2006	2007	2008
Ana	Alex	Arlene	Alberto	Allison	Arthur
Bill	Bonnie	Bret	Beryl	Barry	Bertha
Claudette	Charley	Cindy	Chris	Chantal	Cristobal
Danny	Danielle	Dennis	Debby	Dean	Danny
Erika	Earl	Emily	Ernesto	Erin	Edouard
Fabian	Frances	Franklin	Florence	Felix	Fay
Grace	Gaston	Gert	Gordon	Gabrielle	Gustav
Henri	Hermine	Harvey	Helene	Humberto	Hanna
Isabel	Ivan	Irene	Isaac	Iris	Isidore
Juan	Jeanne	Jose	Joyce	Jerry	Josephine
Kate	Karl	Katrina	Keith	Karen	Kyle
Larry	Lisa	Lee	Leslie	Lorenzo	Lili
Mindy	Matthew	Maria	Michael	Michelle	Marco
Nicholas	Nicole	Nate	Nadine	Noel	Nana
Odette	Otto	Ophelia	Oscar	Olga	Omar
Peter	Paula	Philippe	Patty	Pablo	Paloma
Rose	Richard	Rita	Rafael	Rebekah	Rene
Sam	Shary	Stan	Sandy	Sebastien	Sally
Teresa	Tomas	Tammy	Tony	Tanya	Teddy
Victor	Virginie	Vince	Valerie	Van	Vicky
Wanda	Walter	Wilma	William	Wendy	Wilfred

Eastern North Pacific

Andres	Agatha	Adrian	Aletta	Adolph	Alma
Blanca	Blas	Beatriz	Bud	Barbara	Boris
Carlos	Celia	Calvin	Carlotta	Cosme	Cristina
Dolores	Darby	Dora	Daniel	Dalila	Douglas
Enrique	Estelle	Eugene	Emilia	Erick	Elida
Felicia	Frank	Fernanda	Fabio	Flossie	Fausto
Guillermo	Georgette	Greg	Gilma	Gil	Genevieve
Hilda	Howard	Hilary	Hector	Henriette	Hernan
Ignacio	Isis	Irwin	Ileana	Israel	Iselle
Jimena	Javier	Jova	John	Juliette	Julio

(continues)

Cyclone names (continued)

Kevin	Kay	Kenneth	Kristy	Kiko	Kenna
Linda	Lester	Lidia	Lane	Lorena	Lowell
Marty	Madeline	Max	Miriam	Manuel	Marie
Nora	Newton	Norma	Norman	Narda	Norbert
Olaf	Orlene	Otis	Olivia	Octave	Odile
Patricia	Paine	Pilar	Paul	Priscilla	Polo
Rick	Roslyn	Ramon	Rosa	Raymond	Rachel
Sandra	Seymour	Selma	Sergio	Sonia	Simon
Terry	Tina	Todd	Tara	Tico	Trudy
Vivian	Virgil	Veronica	Vicente	Velma	Vance
Waldo	Winifred	Wiley	Willa	Wallis	Winnie
Xina	Xavier	Xina	Xavier	Xina	Xavier
York	Yolanda	York	Yolanda	York	Yolanda
Zelda	Zeke	Zelda	Zeke	Zelda	Zeke

Central North Pacific

List 1	List 2	List 3	List 4

(The names are used in sequence. When the last name in one list is reached, the next cyclone is given the first name on the next list, regardless of the year. The last name used was Paka [1997], so the next will be Upana.)

Akoni	Aka	Alika	Ana
Ema	Ekeka	Ele	Ela
Hana	Hali	Huko	Halola
Io	Iolana	Ioke	Iune
Keli	Keoni	Kika	Kimo
Lala	Li	Lana	Loke
Moke	Mele	Maka	Malia
Nele	Nona	Neki	Niala
Oka	Oliwa	Oleka	Oko
Peke	Paka	Peni	Pali
Uleki	Upana	Ulia	Ulika
Wila	Wene	Wali	Walaka

Western North Pacific

(There are five lists. Names are used in sequence regardless of the year. Each row of names is contributed by a nation in the region.)

Country	I	II	III	IV	V
Cambodia	Damrey	Kong-rey	Nakri	Krovanh	Sarika
China	Longwang	Yutu	Fengshen	Dujuan	Haima
DPR Korea	Kirogi	Toraji	Kalmaegi	Maemi	Meari
Hong Kong	Kai-Tak	Man-yi	Fung-wong	Choi-wan	Ma-on
Japan	Tenbin	Usagi	Kanmuri	Koppu	Tokage

(continues)

Cyclone names *(continued)*

Lao PDR	Bolaven	Pabuk	Phanfone	Ketsana	Nock-ten
Macau	Chanchu	Wutip	Vongfong	Parma	Muifa
Malaysia	Jelawat	Sepat	Rusa	Melor	Merbok
Micronesia	Ewinlar	Fitow	Sinlaku	Nepartak	Nanmadol
Philippines	Bilis	Danas	Hagupit	Lupit	Talas
Rep. of Korea	Gaemi	Nari	Changmi	Sudal	Noru
Thailand	Prapiroon	Vipa	Megkhla	Nida	Kularb
U.S.A.	Maria	Francisco	Higos	Omais	Roke
Vietnam	Saomai	Lekima	Bavi	Conson	Sonca
Cambodia	Bopha	Krosa	Maysak	Chanthu	Nesat
China	Wukong	Haiyan	Haishen	Dianmu	Haitang
DPR Korea	Sonamu	Podul	Pongsona	Mindule	Nalgae
Hong Kong	Shanshan	Lingling	Yanyan	Tingting	Banyan
Japan	Yagi	Kaziki	Kuzira	Kompasu	Washi
Lao PDR	Xangsane	Faxai	Chan-hom	Namtheun	Matsa
Macau	Bebinca	Vamei	Linfa	Malou	Sanvu
Malaysia	Rumbia	Tapah	Nangka	Meranti	Mawar
Micronesia	Soulik	Mitag	Soudelor	Rananin	Guchol
Philippines	Cimaron	Hagibis	Imbudo	Malakas	Talim
Rep. of Korea	Chebi	Noguri	Koni	Megi	Nabi
Thailand	Durian	Ramasoon	Hanuman	Chaba	Khanun
U.S.A.	Utor	Chataan	Etau	Kodo	Vicete
Vietnam	Trami	Halong	Vamco	Songda	Saola

Western Australia

(The names are used in sequence for all Australian storms. When all the listed names have been used for any part of the country, the same list is used again, starting at the beginning.)

Adeline	Alison	Alex
Bertie	Billy	Bessie
Clare	Cathy	Chris
Daryl	Damien	Dianne
Emma	Elaine	Errol
Floyd	Frederic	Fiona
Glenda	Gwenda	Graham
Hubert	Hamish	Harriet
Isobel	Ilsa	Inigo
Jacob	John	Jana
Kirsty	Kirrily	Ken
Lee	Leon	Linda
Melanie	Marcia	Monty
Nicholas	Norman	Nicky
Ophelia	Olga	Oscar

(continues)

Cyclone names *(continued)*

Pancho	Paul	Phoebe
Rhonda	Rosita	Raymond
Selwyn	Sam	Sally
Tiffany	Taryn	Tim
Victor	Vincent	Vivienne
Zelia	Walter	Willy

Northern Australia

Amelia	Alistair
Bruno	Bonnie
Coral	Craig
Dominic	Debbie
Esther	Evan
Ferdinand	Fay
Gretel	George
Hector	Helen
Jason	Jasmine
Irma	Ira
Kay	Kim
Laurence	Laura
Marian	Matt
Neville	Narelle
Olwyn	Oswald
Phil	Penny
Rachel	Russel
Sid	Sandra
Thelma	Trevor
Vance	Valerie

Eastern Australia

Alfred	Ann	Abigail
Blanch	Bruce	Bernie
Charles	Cecily	Claudia
Denise	Dennis	Des
Ernie	Edna	Erica
Frances	Fergus	Fritz
Greg	Gillian	Grace
Hilda	Harold	Harvey
Ivan	Ita	Ingrid
Joyce	Justin	Jim
Kelvin	Katrina	Kate
Lisa	Les	Larry
Marcus	May	Monica
Nora	Nathan	Nelson
Owen	Olinda	Odette
Polly	Pete	Pierre
Richard	Rona	Rebecca
Sadie	Sandy	Steve
Theodore	Tessi	Tania
Verity	Vaughan	Vernon
Wallace	Wylva	Wendy

Fiji

(Lists A–D are used in sequence, regardless of year. List E is a standby list of replacement names should these be needed.)

A	B	C	D	E
Ami	Arthur	Atu	Alan	Amos
Beni	Becky	Bobby	Bart	Bune
Cilla	Cliff	Cyril	Cora	Chris
Dovi	Daman	Drena	Dani	Daphne
Eseta	Elisa	Evan	Ella	Eva
Fili	Funa	Freda	Frank	Fanny
Gina	Gene	Gavin	Gita	Garry

(continues)

Cyclone names *(continued)*

Heta	Hettie	Helene	Hali	Hagar
Ivy	Innis	Ian	Iris	Irene
Judy	Joni	June	Jo	Julie
Kerry	Ken	Keli	Kim	Koko
Lola	Lin	Lusi	Leo	Louise
Meena	Mick	Martin	Mona	Mike
Nancy	Nisha	Nute	Neil	Nat
Olaf	Oli	Osea	Oma	Odile
Percy	Pat	Pam	Paula	Pami
Rae	Rene	Ron	Rita	Reuben
Sheila	Sarah	Susan	Sam	Solo
Tam	Tomas	Tui	Trina	Tuni
Urmil	Usha	Ursula	Uka	Ula
Vaianu	Vania	Veli	Vicky	Victor
Wati	Wilma	Wes	Walter	Winston
Yani	Yasi	Yali	Yolande	Yalo
Zita	Zaka	Zuman	Zoe	Zena

Papua New Guinea

(The lists are used sequentially.)

A	B
Epi	Abdul
Guba	Emau
Ila	Gule
Kama	Igo
Matere	Kamit
Rowe	Tiogo
Tako	Ume
Upia	

Global warming potential for principal greenhouse gases

Gas	Global warming potential
Carbon dioxide	1
Methane	21
Nitrous oxide	310
CFC-11	3,400
CFC-12	7,100
Perfluorocarbons	7,400
Hydrofluorocarbons	140–11,700
Sulfur hexafluoride	23,900

Ozone depletion potential

Compound	Lifetime (years)	Ozone depletion potential
CFC-11	75	1.0
CFC-12	111	1.0
CFC-113	90	0.8
CFC-114	185	1.0
CFC-115	380	0.6
HCFC-22	20	0.05
Methyl chloroform	6.5	0.10
Carbon tetrachloride	50	1.06
Halon-1211	25	3.0
Halon-1301	110	10.0
Halon-2402	not known	6.0

SI units and conversions

Unit	Quantity	Symbol	Conversion
Base units			
meter	length	m	1 m = 3.2808 inches
kilogram	mass	kg	1 kg = 2.205 pounds
second	time	s	
ampere	electric current	A	
kelvin	thermodynamic temperature	K	1 K = 1°C = 1.8°F
candela	luminous intensity	cd	
mole	amount of substance	mol	
Supplementary units			
radian	plane angle	rad	$\pi/2$ rad = 90°
steradian	solid angle	sr	
Derived units			
coulomb	quantity of electricity	C	
cubic meter	volume	m^3	$1\ m^3 = 1.308\ yards^3$
farad	capacitance	F	
henry	inductance	H	
hertz	frequency	H_z	
joule	energy	J	1 J = 0.2389 calories
kilogram per cubic meter	density	$kg\ m^{-3}$	$1\ kg\ m^{-3} = 0.0624$ lb. ft.$^{-3}$
lumen	luminous flux	lm	
lux	illuminance	lx	
meter per second	speed	$m\ s^{-1}$	$1\ m\ s^{-1} = 3.281$ ft. s^{-1}
meter per second squared	acceleration	$m\ s^{-2}$	
mole per cubic meter	concentration	$mol\ m^{-3}$	
newton	force	N	1 N = 7.218 lb. force
ohm	electric resistance	Ω	
pascal	pressure	Pa	1 Pa = 0.145 lb. in.$^{-2}$

(continues)

SI units and conversions *(continued)*

radian per second	angular velocity	rad s^{-1}	
radian per second squared	angular acceleration	rad s^{-2}	
square meter	area	m^2	1 m^2 = 1.196 yards^2
tesla	magnetic flux density	T	
volt	electromotive force	V	
watt	power	W	1 W = 3.412 Btu h^{-1}
weber	magnetic flux	Wb	

Prefixes used with SI units

Prefixes attached to SI units alter their value.

Prefix	Symbol	Value
atto	a	$\times 10^{-18}$
femto	f	$\times 10^{-15}$
pico	p	$\times 10^{-12}$
nano	n	$\times 10^{-9}$
micro	μ	$\times 10^{-6}$
milli	m	$\times 10^{-3}$
centi	c	$\times 10^{-2}$
deci	d	$\times 10^{-1}$
deca	da	$\times 10$
hecto	h	$\times 10^2$
kilo	k	$\times 10^3$
mega	M	$\times 10^6$
giga	G	$\times 10^9$
tera	T	$\times 10^{12}$

RECOMMENDED READING

Allaby, Michael. *Encyclopedia of Weather and Climate,* 2 vols. New York: Facts On File, 2001.

———. *Ecosystem: Deserts.* New York: Facts On File, 2001.

———. *Guide to Weather.* London: Dorling Kindersley, 2000.

———. *Ecosystem: Temperate Forests.* New York: Facts On File, 1999.

———. *Dangerous Weather: Droughts.* New York: Facts On File, 1998.

———. *Dangerous Weather: Floods.* New York: Facts On File, 1998.

———. *Dangerous Weather: A Chronology of Weather.* New York: Facts On File, 1998.

———. *Dangerous Weather: Tornadoes.* New York: Facts On File, 1997.

———. *Dangerous Weather: Hurricanes.* New York: Facts On File, 1997.

———. *Dangerous Weather: Blizzards.* New York: Facts On File, 1997.

———. *How the Weather Works.* Pleasantville, N.Y: Reader's Digest Association, 1995.

———. *Elements: Fire.* New York: Facts On File, 1993.

———. *Elements: Earth.* New York: Facts On File, 1993.

———. *Elements: Air.* New York: Facts On File, 1992.

———. *A Guide to Gaia.* New York: E.P. Dutton, 1989.

Barry, Roger G., and Chorley, Richard J. *Atmosphere, Weather and Climate,* 7th ed. New York: Routledge, 1998.

Bluestein, Howard B. *Tornado Alley: Monster Storms of the Great Plains.* New York: Oxford University Press, 1999.

Bryant, Edward. *Climate Process and Change.* Cambridge, U.K.: Cambridge University Press, 1997.

Critchfield, Howard J. *General Climatology.* Englewood Cliffs, N.J.: Prentice-Hall, 1960.

Durschmied, Erik. *The Weather Factor: How Nature Has Changed History.* London: Hodder and Stoughton, 2000.

Gleick, James. *Chaos: Making a New Science.* London: Heinemann, 1987.

Henderson-Sellers, Ann, and Robinson, Peter J. *Contemporary Climatology.* London: Longmans, 1986.

Hidore, John J., and Oliver, John E. *Climatology, an Atmospheric Science.* New York: Macmillan, 1993.

Jardine, Lisa. *Ingenious Pursuits: Building the Scientific Revolution.* New York: Doubleday, 1999.

Joseph, Lawrence E. *Gaia: The Growth of an Idea.* New York: St. Martin's Press, 1990.

Lamb, Hubert. *Climate. History and the Modern World,* 2nd ed. New York: Routledge, 1995.

Lovelock, James E. *Homage to Gaia.* Oxford: Oxford University Press, 2000.

————. *The Ages of Gaia.* Oxford: Oxford University Press, 1989.

————. *Gaia: A New Look at Life on Earth.* Oxford: Oxford University Press, 1979.

Lutgens, Frederick K., and Tarbuck, Edward J. *The Atmosphere,* 7th ed. Upper Saddle River, N.J.: Prentice-Hall, 1979.

McIlveen, Robin. *Fundamentals of Weather and Climate.* London: Chapman & Hall, 1992.

MacKenzie, James J., and El-Ashry, Mohamed T. *Air Pollution's Toll on Forests & Crops.* New Haven: Yale University Press, 1989.

Michaels, Patrick J., and Balling, Robert C., Jr. *The Satanic Gases: Clearing the Air about Global Warming.* Washington, D.C.: Cato Institute, 2000.

Oke, T. R. *Boundary Layer Climates,* 2nd ed. New York: Routledge, 1987.

Parkinson, Claire L. *Earth from Above.* Sausalito, Calif.: University Science Books, 1997.

Schneider, Stephen, ed. *Encyclopedia of Climate and Weather,* 2 vols. New York: Oxford University Press, 1996.

Sellers, William D. *Physical Climatology.* Chicago: University of Chicago Press, 1965.

Simons, Paul. *Weird Weather.* London: Little, Brown, 1996.

Singer, S. Fred. *Hot Talk Cold Science: Global Warming's Unfinished Debate.* Oakland, Calif.: Independent Institute, 1997.

Stevens, William K. *The Change in the Weather: People, Weather, and the Science of Climate.* New York: Delacorte Press, 1999.

USEFUL WEBSITES

The Web addresses below refer to particular entries in the Glossary or Biography sections of the *Handbook.* The titles of the relevant entries are arranged alphabetically, with their URLs.

Agassiz: research.amnh.org/ichthyology/neoich/collectors/agassiz.html
www.nceas.ucsb.edu/~alroy/lefa/Lagassiz.html
www.uinta6.k12.wy.us/WWW/MS/8grade/Info%20Access/SPANTLGY/
agassiz.htm

Antarctic circumpolar waves: www.agu.org/pubs/abs/gl/97GL02694/
97GL02694.html

www.ees.lanl.gov/staff/cal/acen.html
jedac.ucsd.edu/wbwhite/ray/paper.html

Arrhenius: www.nobel.se/chemistry/laureates/1903/arrhenius-bio.html

Bernoulli: www-groups.dcs.st-and.ac.uk/~history/Mathematicians/Bernoulli_
 Daniel.html

Celsius: www.astro.uu.se/history/Celsius_eng.html

Coriolis: camille-f.gsfc.nasa.gov/912/geerts/cwx/notes/chap11/gustave.html

Coriolis effect: www.windpower.dk/tour/wres/coriolis.htm

Crutzen: www.nobel.se/chemistry/laureates/1995/crutzen-autobio.html
www.mpch-mainz.mpg.de/~air/crutzen/vita.html

Daniell: www.cinemedia.net/SFCV-RMIT-Annex/rnaughton/DANIELL_BIO.html
www.bioanalytical.com/calendar/97/daniell.htm

dendrochronology: dizzy.library.arizona.edu/library/teams/set/earthsci/
 treering/html
www-personal.umich.edu/~dushanem/whatis.html
www.sonic.net/bristlecone/dendro.html.

FitzRoy: www.intranet.ca/~jedr/fitzroy.htm
www.sciencemuseum.org.uk/collections.exhiblets/weather/fitzroy.htm

Framework Convention on Climate Change: www.unfcc.de/

Galton: www.cimm.jcu.edu.au/hist/stats/galton/index.htm

GISP: gust.sr.unh.edu/GISP2/

GRIP: www.esf.org/life/lp/old/grip/lp_013a.htm

Halley: es.rice.edu/ES/humsoc/Galileo/Catalog/Files/halley.html
www.astro.unibonn.de/~pbrosche/persons/pers_halley.html
www-groups.dcs.st-and.ac.uk/~history/Mathematicians/Halley.html

International Cloud Atlas: orpheus.ucsd.edu/speccoll/weather/27.htm
www.wmo.ch/web/catalogue/New%20HTML/frame/engfil/407.html

Lamb's dust veil index: www.aber.ac.uk/~jpg/volcano/lecture2.html

microburst: www.nssl.noaa.gov/~doswell/microburst/Handbook.html

Molina: www.sciam.com/1197issue/1197profile.html

polar stratospheric clouds: www.atm.ch.cam.ac.uk/tour/psc.html
www.awi-potsdam.de/www-pot/atmo/psc/psc.html

pollen analysis: www.geo.arizona.edu/palynology/plns1295.html

Richardson: maths.paisley.ac.uk/lfr.htm
www.mpae.gwdg.de/EGS/egs_info/richardson.htm

Rowland: fsr10.ps.uci.edu/GROUP/people/drowland.html
www.nobel.se/chemistry/laureates/1995/rowland-autobio.html

Sun photometer: www.concord.org/haze/ref.html

synoptic code: www.met.fsu.edu/Classes/Common/sfc.html
www.zetnet.co.uk/sigs/weather/Met_Codes/codes.html
www.usatoday.com/weather/wpcodes.htm

Tyndall: www.tyndall.uea.ac.uk/tyndall.htm
www.irsa.ie/Resources/Heritage.tyndall.html
www.earthobservatory.nasa.gov/Library/Giants/Tyndall/

Voluntary observing ships: www.vos.noaa.gov/wmo.html
www.srh.noaa.gov/bro/vos.htm

Walker circulation: library.thinkquest.org/20901/overview_2.htm
www.cotf.edu/ete/modules/elnino/cratmosphere.html

Wegener: www.pangaea.org/wegener.htm
www.pbs.org/wgbh/aso/databank/entries/bowege.html
www.ucmp.berkeley.edu/history/wegener.html
www.dkonline.com/science/private/earthquest/contents/hall.html
pubs.usgs.gov/publications/text/wegener.html

INDEX

drought – lifting condensation level